Recent Advancements in the Metallurgical Engineering and Electrodeposition

Edited by Uday Basheer Al-Naib, Dhanasekaran Vikraman and K. Karuppasamy

Published in London, United Kingdom

IntechOpen

Supporting open minds since 2005

Recent Advancements in the Metallurgical Engineering and Electrodeposition
http://dx.doi.org/10.5772/intechopen.77535
Edited by Uday Basheer Al-Naib, Dhanasekaran Vikraman and K. Karuppasamy

Contributors
Eun-Young Choi, Hyun-Chang Oh, Il-Ju Ko, Sukwoo Jung, Safaa Najah Saud, Tapan Sarkar, Uday M. Basheer Al-Naib, Carlos Roberto Grandini, Daniela Cascadan, Oleksandr Salenko, Ayodeji Afolabi, Api Popoola, Boukhlifi Fatima, Jianxun Song, Yusi Che, Yongchun Shu, Jilin He, Venkata Saravanan K, Vineetha P

Notice
Statements and opinions expressed in the chapters are these of the individual contributors and not necessarily those of the editors or publisher. No responsibility is accepted for the accuracy of information contained in the published chapters. The publisher assumes no responsibility for any damage or injury to persons or property arising out of the use of any materials, instructions, methods or ideas contained in the book.

First published in London, United Kingdom, 2020 by IntechOpen
IntechOpen is the global imprint of INTECHOPEN LIMITED, registered in England and Wales, registration number: 11086078, 7th floor, 10 Lower Thames Street, London, EC3R 6AF, United Kingdom
Printed in Croatia

British Library Cataloguing-in-Publication Data
A catalogue record for this book is available from the British Library

Additional hard and PDF copies can be obtained from orders@intechopen.com

Recent Advancements in the Metallurgical Engineering and Electrodeposition
Edited by Uday Basheer Al-Naib, Dhanasekaran Vikraman and K. Karuppasamy
p. cm.
Print ISBN 978-1-78984-686-7
Online ISBN 978-1-78984-687-4
eBook (PDF) ISBN 978-1-78985-908-9

For EU product safety concerns:
IN TECH d.o.o., Prolaz Marije Krucifikse Kozulić 3, 51000 Rijeka, Croatia, info@intechopen.com or visit our website at intechopen.com.

Meet the editors

Dr. Uday M. Basheer Al-Naib is a senior lecturer at Universiti Teknologi Malaysia. He completed a postdoctorate in Materials Engineering at Universiti Sains Malaysia (USM) between 2013 and 2014. He received his PhD in Materials Engineering from USM in 2013 and obtained his master's and bachelor's degrees in Chemical Engineering from the University of Baghdad and University of Technology, Iraq, respectively. He has over 30 years of academic and practical experience in the field of ceramic and metal composites, metal alloys, advanced ceramics, and ceramic–metal joining. He is the editor of a book series from IntechOpen. In addition, he is a recognized reviewer for Elsevier. He is acknowledged at the practical level for solving specific problems related to materials industries. This was evident when he published his research work with different publishers and in international material engineering journals with high impact factors. In addition to his own research, Dr. Uday has been acting as a supervisor of several academic theses related to different fields of materials engineering. He received his CEng from Engineering Council, UK, and became a professional members of the Welding Institute in May 2019. He is also a professional member (ProfGradIMMM) of the IOM3 and a fellow member (FIMM) of the Institute of Materials, Malaysia.

Dr. Dhanasekaran Vikraman is an assistant professor in the Division of Electronics and Electrical Engineering, Dongguk University-Seoul, Seoul, Korea. He received his bachelor's degree at Mannai Rajagopalaswamy Govt. Arts College, Mannargudi, affiliated to Bharathidasan University, Tiruchirappalli, India. After that, he completed his master's and PhD degrees at the Department of Physics, Alagappa University, India. Later, he received a visiting scientist position at KIST, Korea; a Marie-Curie Experienced Researcher fellowship at the Department of Physics, Aristotle University of Thessaloniki, Greece; and a postdoctorate position at Sejong University and Ajou University, Korea. He has authored more than 120 international journal articles and two book chapters. In addition, he served as an editor with IntechOpen for a book entitled *Design, Simulation and Construction of Field Effect Transistors*.

Dr. K. Karuppasamy is currently working as an assistant professor at the Division of Electronics & Electrical Engineering, Dongguk University-Seoul, South Korea. He obtained his master's degree in Chemistry from VHNSN College, Tamil Nadu, India, in 2010. He received his PhD in the field of Nanocomposite Electrolytes for Lithium-Ion Batteries from Anna University, Chennai, in 2014. Later, he completed his postdoctorate positions at the Department of Metallurgical Engineering and Materials Science, Indian Institute of Technology, Bombay, India, and the Department of Chemical Engineering, Sogang University, Seoul, South Korea. His research interests mainly focus on nanostructured materials for energy storage and conversion devices. He has published more than 60 research papers in international journals and presented more than 25 papers in international and national conferences.

Contents

Preface

Metallurgical engineering and electrodeposition of metals and alloys are of great importance in several fields such as physical metallurgy, extractive metallurgy, mineral processing electroplating, electrowinning, and electrorefining. This book shares recent research and knowledge related to metallurgical engineering and electrodeposition, and contains 10 chapters. Chapter 1 briefly explains the introduction of engineering materials and metallurgy. Chapter 2 introduces the importance of understanding the role of substitutional nickel in the structure, microstructure, and selected mechanical properties of titanium–nickel (Ti–Ni) system alloys. Chapter 3 shows the effect of alloying elements such as Ti, Mn, Be, Zr, and B on the phase transitions and enhancement of the mechanical properties of Cu-based shape memory alloys. Chapter 4 discusses the fabrication of dielectric thick film by electrophoretic deposition and its characterization. Chapter 5 illustrates the general principles and electrolyte optimization of cooperative interaction mechanisms between metastable-state ions and electrolytes.

Chapter 6 explains the metal organic framework of ionic liquids. Chapter 7 presents the characterization and treatment of real wastewater from an electroplating company using raw chitin. Chapter 8 deals with the mechanical properties analysis of austempered ductile iron weld joints using a developed electrode. Chapter 9 exposes the influence of spark plasma sintering parameters on the mechanical properties of high-entropy alloys for aerospace applications. Chapter 10 explains the cutting of super-hard materials using jet methods. The purpose of this chapter is to develop the principles of a functional approach to create hybrid processes in a morphological combination of energy flows impacted on a workpiece.

We hope that the knowledge and objectives of recent research achievements in metallurgical engineering and electrodeposition will be useful for researchers, students, engineers, and others that work with these kinds of materials.

I acknowledge IntechOpen for their enthusiastic and professional support. I would like also to express my thanks to the editor panel, Dr. Dhanasekaran Vikraman and Dr. K. Karuppasamy, for their endless scientific support, and to the authors who contributed their findings related to metallurgical engineering and electrodeposition. Without their expertise and contributions, this book would not have been possible.

Dr. Uday M. Basheer Al-Naib
School of Mechanical Engineering,
Faculty of Engineering,
Centre for Advanced Composite Materials (CACM),
Institute for Vehicle Systems and Engineering,
Universiti Teknologi Malaysia,
Skudai, Johor, Malaysia

Dr. Dhanasekaran Vikraman and Dr. K. Karuppasamy
Division of Electronics and Electrical Engineering,
Dongguk University-Seoul,
Seoul, South Korea

Chapter 1

Introductory Chapter: A Brief Introduction to Engineering Materials and Metallurgy

Uday M. Basheer Al-Naib

1. Introduction

Materials science, also known as materials science and engineering, is a multidisciplinary field that deals with the discovery and design of new materials [1]. Always new materials open the door to new technologies, whether they are in chemical, civil, construction, nuclear, agricultural, aeronautical, biomedical, electrical, or mechanical engineering [2, 3]. The materials science engineering includes the study of the relations between the synthesis, forming, properties, structure, and performance of materials that enable an engineering function. The materials properties of interest can be electrical, mechanical, optical or magnetic; the engineering function can affect industries involved in electronics, communications, transportation, manufacturing, medicine, recreation, environment, and energy [4].

Metallurgy is the art and science of making metals and alloys in shapes and with characteristics suitable for practical use [5]. It is an applied science based on a clear understanding of the structures and properties of metals and their alloys. Metallurgy has long occupied the dominant position as the most important engineering materials; steel being by far the most important over the last few centuries. However, increasingly in many areas other materials such as ceramics, plastics and composites are challenging this position. The relatively recent development of materials science and engineering is a merger of metallurgy with others like glass and ceramic technology, mineralogy, physical and inorganic chemistry, solid-state physics, and polymer science in modern techniques to include all structural and functional materials, thus making it one of the widest of study disciplines [1].

2. What is metallurgy?

Metallurgy is the science and technology to extract metals from their ores economically, refining them and preparing them for the end use [6]. It studies the microstructure of a metal, the structural features that are control to observation under a microscope. Microstructure determines mechanical properties of the metal, including their elastic and plastic behavior when applying the force. Chemical composition is the relative content of a particular element within an alloy, usually expressed as a percent weight. Composition, as well as thermal and mechanical processing, will determine microstructure. Metals and their alloys are widely used in our daily lives. They are used for different purposes such as making machines,

IntechOpen

bridges, motor vehicles, railways, buildings structure, ships, aircrafts, agricultural tools, etc. Therefore, real economic growth can come from increasing quality and quantity of the metal production in that country [7].

Naturally, most metals occur in the combined state as minerals and they are reactive. Only a few metals like gold, silver, platinum and mercury, etc. are found as Free State in the earth's crust. Metals that have a low reaction show little convergence to air, moisture, carbon dioxide, or non-metals found in nature [8]. Materials that occurring naturally in which a metal or its compound occurs is called a mineral. A mineral from which a metal can be economically extracted is called an ore. The main active components found in nature, especially in the atmosphere are oxygen and carbon dioxide. In the earth's crust, silicon and sulfur are present in large quantities. Seawater also contains large amounts of chloride ions (obtained from dissolved salts). Most active metals are high electrically positive and therefore exist as different ions [9]. For this reason, most of the important ores of these metals occur as different components such as oxides, silicates, carbonates, and halides.

3. Metallurgy forming and processing

The importance of metals in modern technology is largely due to the ease with which it can be formed in useful shapes [10]. Hundreds of processes have been developed for specific applications of metalworking. However, these processes can be categorized into only a few classes based on the type of force applied to the workpiece when it is formed [11]. These classes are direct-compression-type processes, Indirect-compression processes, shearing processes, bending processes, and tension-type processes as shown in **Figure 1**. In direct compression processes, force is applied to the workpiece surface, and the metal flows at an angle based on the pressure direction. In indirect-compression processes, the basic applied forces are often tensile, but the indirect compressive forces are developed by the reaction of the workpiece with the die up to high values. These processes include extrusion, pipes, deep drawing of the cup and pulling wires. Therefore, the metal flows under the influence of a combined stress condition involving high pressure forces in at least one of the main directions. The best example of a tension-type forming process is the formation of expansion, where the sheet of metal is wrapped in a die contour under tensile forces. Shearing involves applying the shearing forces of sufficient size to tear the metal in the plane of shear, while bending involves applying the bending moments on the metal sheet. **Figure 1** shows these processes in a very basic way.

Metallurgy forming processes of are usually classified into hot and cold working processes. Hot working is defined as deformation under temperature and strain rate conditions so that recovery operations are performed together with the deformation. On the other hand, cold working is deformed in circumstances where recovery operations are not effective [12]. In hot working, the strain hardening and deformed grain structure caused by deformation are quickly eliminated by the formation of new strain-free grains as the result of recrystallization and grain growth. It is possible to have very large deformities in hot working because the recovery processes keep up with deformation [13]. Hot working occurs when the flow stress is essentially constant. The energy required for deformation is generally lower for hot working compared to cold working because of the flow stress decreases with increasing temperature. Since strain hardening is not alleviated in cold working, the flow stress increases with increasing the deformation. Therefore, the total plastic deformation without fracture is less for cold working compared with hot

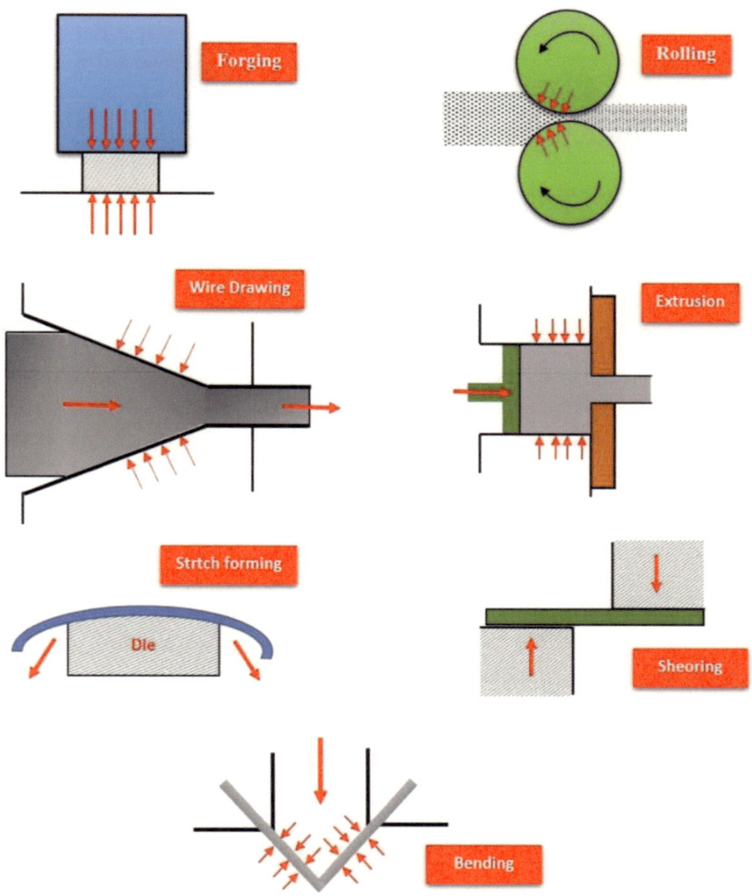

Figure 1.
Typical metallurgy forming operations.

working, unless the effects of cold work are mitigated through annealing process. It is important to understand that the difference between cold working and hot working does not depend upon any arbitrary deformation temperature. For most commercial metal alloys, hot working process should be performed at a relatively high temperature in order to obtain a rapid recrystallization rate. However, lead and tin recrystallize rapidly at room temperature after significant deformations so that the working of these metals at room temperature is like hot working. Similarly, the work of tungsten at 1093°C, in the hot work range of the steel, is a cold work because this high melting metal has a recrystallizing temperature higher than this working temperature (**Figure 1**) [14].

4. Mechanical testing and materials characterizations

Whether a material is suitable for a given application is specified by the material properties. These properties can be measured using a series of mechanical tests, such as tensile, compressive, hardness and fatigue testing (**Figure 2**) as well as physical and chemical tests. Some of the mechanical tests are easily accessible like hardness. Others are difficult to measure such as tensile or yield strength where special samples must be formed. It is difficult to determine other properties such as fatigue, toughness strength as the tests need several

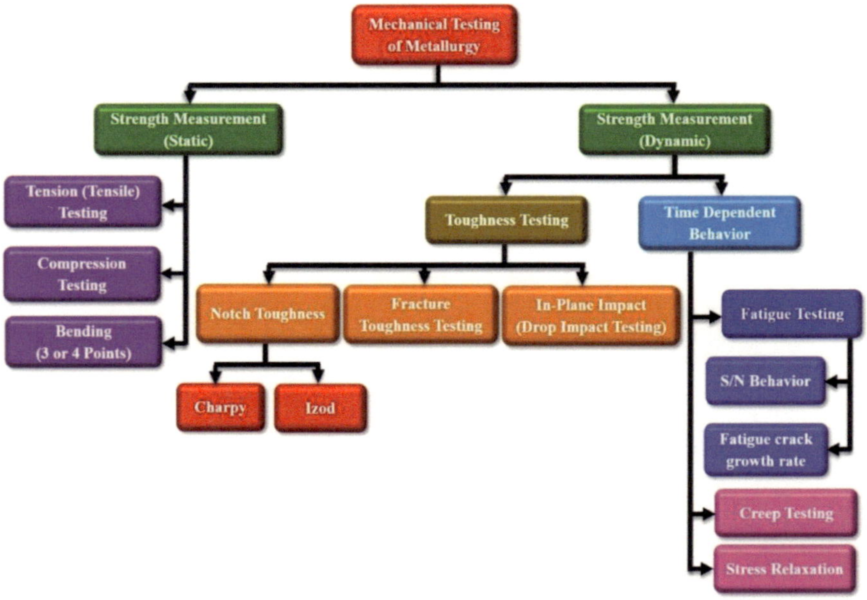

Figure 2.
Commonly used mechanical testing techniques.

samples per every case and the testing process takes a long time. Apart from the above tests, it is also possible to predict the properties of materials by determining the microscopic structure of materials, where properties are determined by the microstructure. There are a number of microstructural characterization techniques including optical microscopy, scanning electron microscopy (SEM), electron backscattered diffraction (EBSD), X-ray diffraction (XRD), transmission electron microscopy (TEM), ultrasonic sound based methods and magnetic-based methods [15]. Among these techniques, optical microscopy and electron microscopy are able to detect the morphology of microstructural features in the surface of a prepared sample. Optical microscopy has its own advantages such as low cost, ease of use on large sample areas and ease of operation. However, electron microscopy is also widely used due to its high resolution down to the nanometer scale [16].

The factors that often determine the properties of strength and toughness of pearlitic steels like interlamellar spacing, colony size, and prior austenite size can be done by examining the microstructure of these materials [17, 18]. However, measuring the size of the colony and especially the size of the prior-austenite grains is extremely difficult and requires a proficient technical examination using optical microscopy or SEM and special procedures. For multiple phases of steels, the fraction, morphology, size and distribution of the phase components are determined by the properties. XRD is an effective technique for measuring the fraction of the present phases, but cannot access the size, morphology and distribution. Image analysis techniques are also applied for this application. Unlike the XRD analysis, which is a crystallographic analysis of the bulk surface, image analysis technique extracts information directly from a microscopic image of the sample surface. So once the current features in the sample images are categorized into phases, the size, fraction, morphology and phase's distribution can easily be obtained. Most of the current image analysis-based characterization uses the histogram of the brightness (intensity) of the individual pixels that make up the image, and relies on all the pixels in one phase having intensities in a

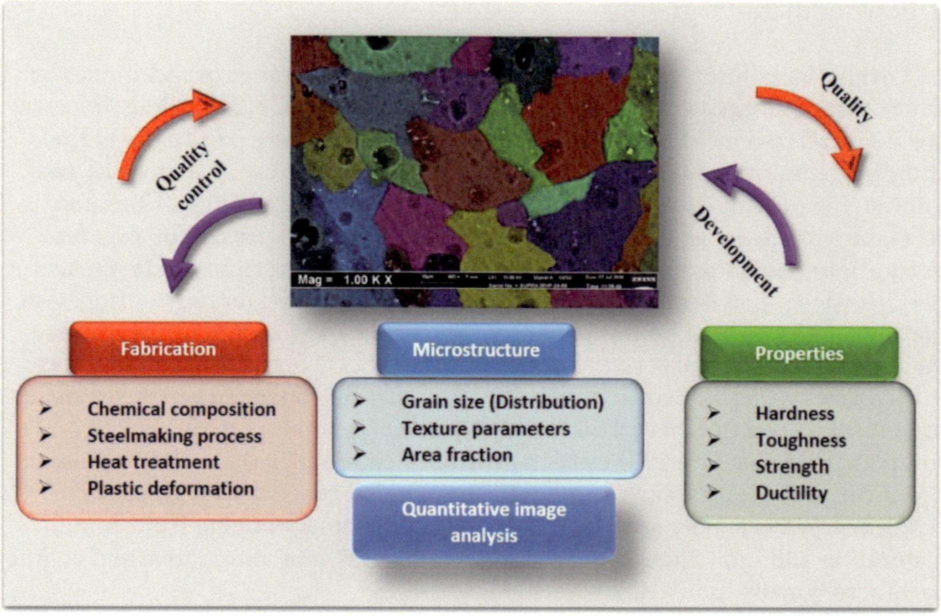

Figure 3.
Microstructure related with fabrication of components and their properties.

different range from all of those in another phase [19]. This makes it very easy to distinguish between the phases only using the threshold, and it has been shown to work with some two-phase steels characterized by high variability between phases [20]. However, the different phases in brightness levels overlap for many other steels with complex microscopic structures. In this case, the threshold of intensity is no longer able to distinguish between the phases, and instead some analysis of the spatial patterns of intensity within the phases, i.e., "texture," is required. Quantitative analysis of microstructural images allows not only a quality control examination of the treatment path, but also the possibility of establishing a reciprocal relationship between microstructure features and associated properties (**Figure 3**).

5. Steel metallurgy

Steel has been one of the most important materials used by humans for up to 4000 years due to its good combination of low cost and properties. The mechanical properties of steel have been found to be highly dependent on its internal structure at nanometers up to microns or even millimeters (its "microstructure"). The internal structure of the steel can be adjusted through composition changes, mechanical deformation or heat treatments. The metal can then be designed to meet the different requirements in a range of applications. Atoms can be arranged into steel and bonded, called phases in several ways. Different phases have various properties, which may be suitable for various applications, either singly or in groups. For example, high-strength steels have good wear and abrasion resistance as well as high tensile strength, so they are widely used as rail steels and steel wire [21]. Steel with more than one phase including two-phase steels, complex steels and transformation induced plasticity steels usually has a good mix of toughness and strength and is therefore well suited for applications that require strength and formability as in the automotive industry [22].

6. Powder metallurgy

Powder metallurgy (PM) is a term that covers a wide range of techniques by which materials or components are made of metal powders [23]. These processes are characterized by high productivity and ideal for making parts close to the complex geometry of a range of materials, increasing the use of materials, and minimizing or eliminating secondary processes such as machining. Common secondary processes of components made by liquid metal processing may lead to additional manufacturing steps that have significant cost and waste impacts [24]. However, powder metallurgy processing has obtained more attention with the advantages offered by comparison with casting and forging. In PM processing techniques, all or part of the some constituents are formed by compressing the particles with the characteristic structure, size and shape in a high-precision product. The ability of metal metallurgy powders to produce high-quality, complex parts characterized by high productivity and high durability represents significant advantages, such as potential low capital costs, with high energy efficiency. PM is widely used for a range of applications, ranging from dental restoration and implantation to bearings, transmission of automotive and engine parts across many industrial sectors (**Figure 4**).

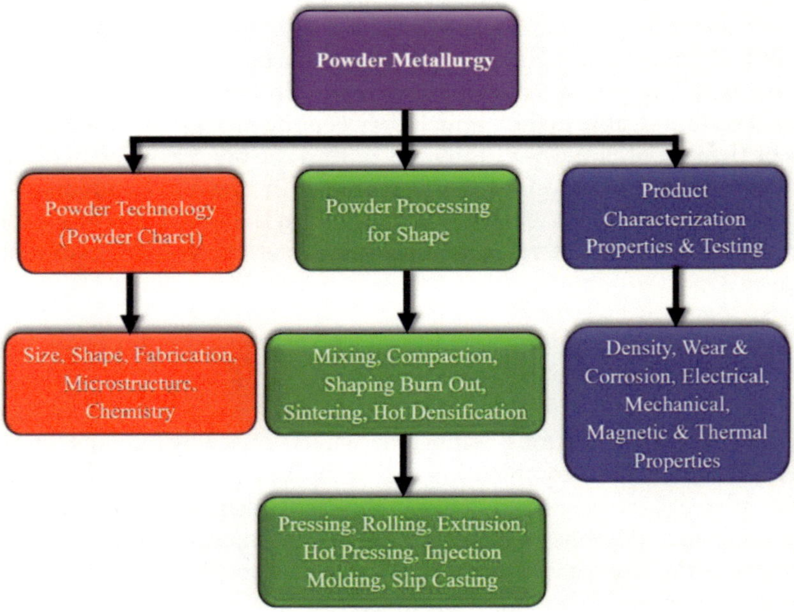

Figure 4.
Schematic of powder metallurgy.

Author details

Uday M. Basheer Al-Naib
Faculty of Engineering, School of Mechanical Engineering, Universiti Teknologi Malaysia, Malaysia

*Address all correspondence to: ummb2008@gmail.com

References

[1] Ferguson C. Historical Introduction to the Development of Material Science and Engineering as a teaching discipline. UKCME; 2006. Available from: http://www.materials.ac.uk/pub/Materials-History-lntro. pdf

[2] Tobins FHA, Oladiran KA, Muriana RA, Abdulrahman AS. Snail Shell as an inspiring engineering material in science and technology development: A review. International Journal of Contemporary Research and Review. 2018;**9**(03):20408-20416

[3] Cheshier SR. The field of engineering technology. In: Studying Engineering Technology: A Blueprint for Success. Chapter 2. 2006:1-45

[4] Fine MEM, Harris L. Materials science and engineering, an educational discipline. Annual Review of Materials Science. 1994;**24**(1):1-19

[5] Reuter MA, Antoinette VS, Jens G, Neill B, Abadías-Llamas A. Challenges of the circular economy: A material, metallurgical, and product design perspective. Annual Review of Materials Research. 2019;**49**:1-49

[6] Gordon R. Transformative innovation in mining and metallurgy. In: Metallurgical Design and Industry. Springer; 2018;**2**:155-191

[7] Braga B, Lerman RI. Accounting for homeownership in estimating real income growth. Economics Letters. 2019;**174**:9-12

[8] Choi SS, I B, Hong J, Yoon KJ, Son JW, Lee JH, et al. Catalytic behavior of metal catalysts in high-temperature RWGS reaction: In-situ FT-IR experiments and first-principles calculations. Scientific Reports. 2017;**7**:41207

[9] Bi W, Xiaogang L, You R, Chen M, Yuan R, Huang W, et al. Surface immobilization of transition metal ions on nitrogen-doped graphene realizing high-efficient and selective CO_2 reduction. Advanced Materials. 2018;**30**(18):1706617

[10] Tofail SAK, Elias P, Bandyopadhyay A, Bose S, O'Donoghue L, Charitidis C. Additive manufacturing: Scientific and technological challenges, market uptake and opportunities. Materials Today. 2018;**21**(1):22-37

[11] Poddar P. Principles of Metal Working Processes. 2008:64-77

[12] McQueen H, Jonas J. Recovery and recrystallization during high temperature deformation. Treatise on Materials Science & Technology. 1975;**6**:393-493

[13] Sakai T, Andrey B, Kaibyshev R, Miura H, Jonas JJ. Dynamic and post-dynamic recrystallization under hot, cold and severe plastic deformation conditions. Progress in Materials Science. 2014;**60**:130-207

[14] Kang J-Y, Hoyoung K, Son D, Kim C, Park SK, Lee T-H. Hot-worked microstructure and hot workability of cold-work tool steels. Materials Characterization. 2018;**135**:8-17

[15] Aguiar IV, Pérez ED, Santos DB, Modenesi PJ. Microstructure characterization of a duplex stainless steel weld by electron backscattering diffraction and orientation imaging microscopy techniques. Matéria (Rio de Janeiro). 2015;**20**(1):212-226

[16] Lin P-C, Stephen L, Wang PC, Sridhar R. Techniques for physicochemical characterization of nanomaterials. Biotechnology Advances. 2014;**32**(4):711-726

[17] Hyzak J, Bernstein I. The role of microstructure on the strength

and toughness of fully pearlitic steels. Metallurgical Transactions A. 1976;**7**(8):1217-1224

[18] Bouse G, Bernstein I, Stone D. Role of alloying and microstructure on the strength and toughness of experimental rail steels. In: Rail Steels—Developments, Processing, and Use. Stone D, Knupp G.(editors) West Conshohocken, PA: ASTM International; 1978:45-161. https://doi.org/10.1520/STP27107S

[19] Russ JC. The Image Processing Handbook. CRC Press; 2016

[20] Burger W, Burge MJ. Digital Image Processing: An Algorithmic Introduction Using Java. Springer; 2016

[21] Cutler CP. Use of metals in our society. In: Metal Allergy. Springer; 2018. pp. 3-16

[22] Totten GE. Steel Heat Treatment: Metallurgy and Technologies. CRC Press; 2006

[23] German RM. Powder Metallurgy and Particulate Materials Processing: The Processes, Materials, Products, Properties, and Applications. Princeton: Metal Powder Industries Federation; 2005

[24] Miracle D. Metal matrix composites—From science to technological significance. Composites Science and Technology. 2005;**65**(15-16):2526-2540

Chapter 2

Structure, Microstructure, and Some Selected Mechanical Properties of Ti-Ni Alloys

Daniela Cascadan and Carlos Roberto Grandini

Abstract

Titanium nickel (Ti-Ni) alloys with low nickel (Ni) content can be used as biomaterials because they improve the mechanical properties, corrosion, and wear resistance of commercially pure titanium (Cp-Ti). Among the mechanical properties of a biomaterial, elastic modulus and microhardness are very important, and each varies according to the microstructure and interstitial elements such as oxygen and nitrogen as well as the amount of substitutional element and thermomechanical processing. Heat treatments are used to obtain a homogeneous microstructure, free of internal stresses structural, microstructural, also to retain or change the size of the phases. In this chapter, the preparation, chemical, structural, and microstructural, and mechanical characterization of Ti-Ni alloys are presented. The structural and microstructural characterization showed the predominant presence of α and Ti_2Ni phases. There is no clear variation of the microhardness due to the amount of nickel. The dynamic elastic modulus was slightly above the Cp-Ti due to the addition of a new intermetallic phase (Ti_2Ni) but did not vary significantly with the amount of Ni.

Keywords: Ti-Ni alloys, structure, microstructure, mechanical properties

1. Introduction

The development of biomaterials has created several significant benefits for the general population over the last few decades, including dental implants, prosthetics, artificial arteries, and contact lenses [1]. These benefits have been either for the purpose of correcting problems or have been esthetic in nature [2].

Various materials have been used as biomaterials. Titanium alloys [3, 4] are among these materials because they have excellent mechanical strength/density ratio, excellent corrosion resistance, and biocompatibility.

The mechanical properties [5], corrosion, and wear resistance [6] of a material are largely dictated by the microstructure. Titanium alloys are favorable because a wide spectrum of microstructures can be obtained, depending on the chemical composition and the processing. This makes titanium alloys advantageous because they allow the desired microstructure to be obtained for specific requirements, as low elastic modulus, for example, [7, 8].

Titanium exists in two allotropic forms. At low temperatures, the so-called alpha phase (with hexagonal compact crystalline structure) is presented, and the beta phase (with body-centered cubic crystalline structure) emerges above 883°C. Some elements called beta stabilizers, such as niobium, molybdenum, iron, vanadium, and nickel, lead to a decrease in beta transus temperature (the transition from alpha to beta phase) when added to the forming titanium alloys and may stabilize this phase at room temperature. These phases determine a classification of titanium alloys: alloys containing only α stabilizers at this phase are known as α alloys, alloys containing 1–2% beta stabilizers and about 5–10% of beta phase are called near-α, alloys containing increased amounts of beta stabilizers resulting in 10–30% of beta phase are called α + β alloys, and alloys containing increased amounts of beta stabilizers where this phase can be retained by quick cooling are known as β metastable phases (these alloys decay into α + β in aging treatments) [9]. Beta transus temperature fulfills a central role in the microstructure's evolution, and so it is of great importance to determine the type of technological processing and heat treatment, which also includes doping with oxygen or nitrogen, whose elements present in the crystalline lattice can significantly affect the properties of the same [10].

In addition to interfering significantly in the structure and microstructure of titanium alloys, substitutional elements interfere in mechanical properties such as elastic modulus and hardness to the extent that they stabilize phases with different properties [11, 12]. The α phase, with a greater packaging factor of atoms, has a greater elastic modulus in relation to phase β, due to greater proximity between the atoms. Regarding toughness, it is usually greater in α phase, which contains fewer plans of slipping in relation to β phase, making the material harder and brittle [13]. However, depending on the alloy and the processing performed, alloys with a β structure have similar hardness [14].

The presence of atoms in interstitial positions of the lattice (e.g., nitrogen and oxygen) strongly influences some properties, such as mechanical resistance and ductility [15]. These elements act as stabilizers in titanium's α and α' phases, increasing the temperature of β-transus [16, 17].

Nickel causes several adverse reactions in the human body, such as carcinogenic, genotoxic, mutagenic, cytotoxic, and allergenic effects [18]. However, when nickel is connected to titanium, these problems are minimized [19]. In a study of corrosion of titanium alloys with varied concentrations of nickel in Ringer's solution, it was found that the potential of disruption occurs from 29 wt% of nickel and decreases rapidly, indicating that alloys below this concentration present promising applications regarding corrosion [20, 21]. A later study on Hank's biocorrosion solution concluded that the daily release of nickel would be hundreds of times smaller than that contained in water intake [22]. In addition, the microhardness of these alloys is similar to enamel (310–390 HVN), which is suitable for use in dental screws [18].

Thus, it is of fundamental importance to understand the role of substitutional nickel in the structure, microstructure, and selected mechanical properties in Ti-Ni system alloys. To this purpose, Ti-Ni alloys containing 5, 10, 15, and 20 wt% of nickel were melted in an arc-melting furnace, mechanically processed by hot rolling, and subjected to heat treatment in a vacuum to cause homogenization and stress relief.

2. Materials and methods

2.1 Ti-Ni alloys

The phase diagram of the Ti-Ni system is presented in **Figure 1**. This phase diagram shows that alloys considered in this chapter (5, 10, 15, and 20 wt% of nickel) present the phases α and β (both of titanium) and the intermetallic phase Ti_2Ni, at high temperatures [23]. By the lever rule, the higher the nickel

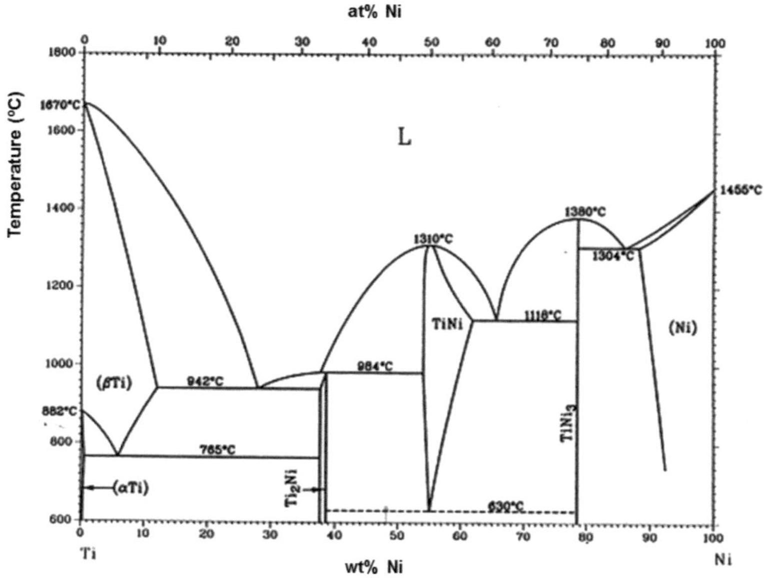

Figure 1.
Phase diagram of the Ti-Ni system [23].

concentration is, the greater the amount of this intermetallic is. By the decay of the β-transus curve, it is concluded that the addition of nickel favors the stabilization of this phase. The eutectoid reaction $\beta \rightarrow \alpha + Ti_2Ni$ occurs at 765°C. By the lever rule, in the field $\beta + Ti_2Ni$, the amount of Ti_2Ni is lower in relation to $\alpha + Ti_2Ni$ field, in the same concentration.

For example, for an alloy with 10 wt% nickel that underwent a homogenization heat treatment at 1000°C, followed by slow cooling, in equilibrium conditions, at this temperature, the alloy presents the β phase of titanium. With cooling, it begins to precipitate the intermetallic phase Ti_2Ni. At a temperature just above 765°C, the precipitates of Ti_2Ni proeutectoid phase are found in a β matrix of eutectoid composition and just below 765°C and are found in the eutectoid matrix, with a granular microstructure of α and Ti_2Ni phases intercalated. From there, the intermetallic Ti_2Ni composes both precipitates and the eutectoid matrix.

Theoretically, the phases present in these alloys at room temperature are $\alpha + Ti_2Ni$; however, when the eutectoid reaction is suppressed, as in the case where the cooling is not done in perfect equilibrium condition, there is the possibility of β-phase retention at room temperature, the higher the nickel concentration is [24].

According to Ti-Ni binary diagram, a eutectic $L \rightarrow \beta + Ti_2Ni$ to 942°C occurs at a concentration of 28.4 wt% of nickel, <700°C in relation to the melting point of pure titanium (1670°C), thus decreasing the melting point and conforming temperature. Therefore, in the case of alloys with 15 and 20 wt% of nickel, at some point during the melting, there is the possibility of developing a microstructure composed of pre-eutectic precipitates of the β phase in a eutectic matrix, although the melting is not done under equilibrium conditions.

The Ti-Ni alloys were prepared using commercially pure titanium (Cp-Ti) (99.7% purity, Aldrich Inc., USA) and nickel (99.5% purity, Camacam Industrial, Brazil). The ingots were obtained by arc melting, using a water-cooled copper crucible and an argon-inert atmosphere. The melting was held five times to ensure the homogeneity of the produced ingots. Samples were prepared with 5, 10, 15, and 20 wt% of nickel. Chemical analysis of the samples was performed using an inductively coupled plasma optical emission spectrometry (ICP-OES), Varian, Visa model, from the sample's dissolution in acid medium. The presence of interstitial elements was carried out using a LECO TC-400 model gas analyzer. **Table 1** shows the chemical analysis of the produced samples.

It can be verified that the obtained ingots show the correct stoichiometry of nickel and a low amount of impurities and interstitial elements, displaying the good quality of prepared alloys.

To check the homogeneity of the produced alloys, the samples were subjected to a mapping of the elements that comprise the alloys using an electron dispersion spectrometry (EDS) technique that used an Oxford Instruments Incax-act model detector coupled to the scanning electron microscope (SEM) Carl Zeiss EVO/LS15 model. **Figure 2** shows the mapping of the elements titanium and nickel obtained by EDS in the produced alloys after melting. In all cases, homogeneous distribution of the elements titanium and nickel can be observed, showing that the melting was complete, without the presence of aggregates or segregates.

After melting, the ingots were submitted to a homogenization heat treatment that consisted of slowly heating the samples up to 1173 K for 24 h in a vacuum of 10^{-6} Torr, followed by slow cooling in the furnace. The mechanical processing of

Alloy	Fe (wt%)	Al (wt%)	Ni (wt%)	Mn (wt%)	Cr (wt%)	Cu (wt%)	C (wt%)	O (wt%)	N (wt%)
Ti-5Ni	0.07	0.005	5.36	0.020	0.008	0.001	0.124	0.109	0.012
Ti-10Ni	0.10	0.018	10.03	0.034	<0.001	<0.001	0.380	0.179	0.007
Ti-15Ni	0.12	0.035	15.83	0.052	<0.001	<0.001	0.023	0.290	0.058
Ti-20Ni	0.12	0.028	19.80	0.072	<0.001	<0.001	0.160	0.300	0.076

Table 1.
Chemical analysis of the Ti-Ni alloys.

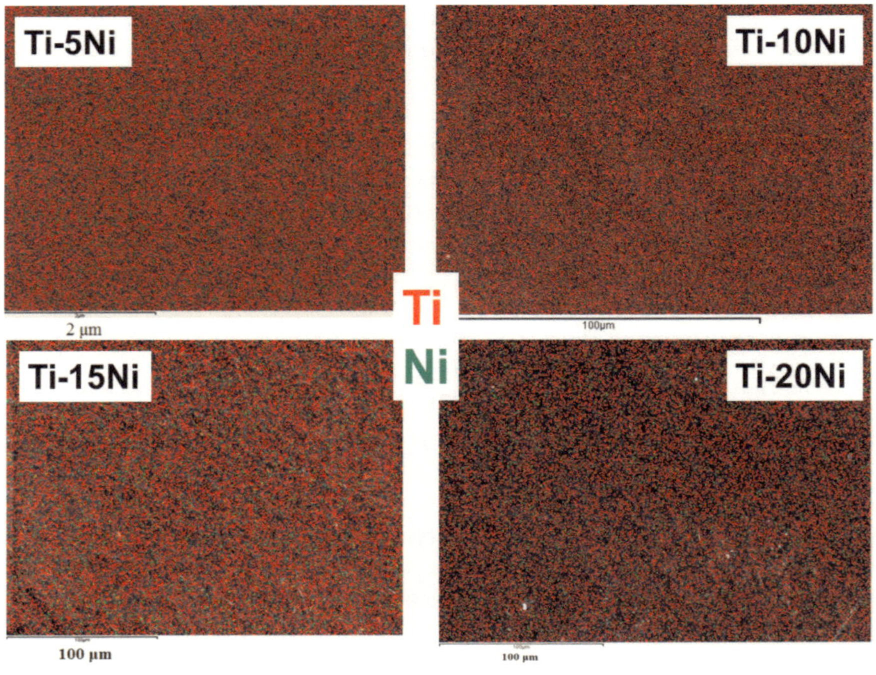

Figure 2.
EDS mapping of the elements Ti and Ni that make up the Ti-Ni alloys.

the samples is important because it allows tests that require symmetrical samples, in addition to causing changes in microstructure and some properties that are of interest in the analysis. The samples were hot rolled as their mechanical processing.

After slow cooling, a new heat treatment was performed for strain relief and recrystallization of microstructures as there were internal stresses and deformed microstructures due to an aggressive hot-rolling process. This treatment was performed at 1143 K for 24 h in a vacuum.

2.2 Sample's characterization

The structural characterization of the samples was carried out by X-ray diffraction (XRD) measurements on a Rigaku diffractometer (D/Max-2100PC model). This included a Cu-Kα radiation of wavelength of 1.544 Å, 20 mA current, and a potential of 40 kV. The fixed mode, with step of 0.02° and time of 1.6 s, in the range of 10–90°, was used. The used method to data acquisition was the powder method [25].

For all samples of the Ti-Ni alloys used in this chapter and in all processing conditions, the microstructure was evaluated by both optical and scanning electron microscopy. To obtain the images, an Olympus model BX51M optical microscope with a camera connected to a computer and a SEM Carl Zeiss EVO/LS15 model with the SmartSEM software were used.

Vickers microhardness tests were performed using a Shimadzu HMV model-2 microdurometer connected to a computer. The load used was 1.941 N for a time of 60 s. Following ASTM standards, almost 20 indentations were made in different regions of the sample [26, 27]. The dynamic elastic modulus was measured using Sonelastic® equipment (ATCP, Brazil). A boost in sample, the frequencies of vibration (through the Sonelastic® software), and damping were calculated and linked with the modulus of elasticity, according to standard ASTM E1876-01 [28].

3. Results and discussion

3.1 Structural characterization

Figure 3 shows the X-ray diffractograms of all samples of Ti-Ni alloys after melting. During analysis of the X-ray diffractograms of samples after melting, it was observed that the addition of at least 10% in the weight of nickel caused the appearance of other phases beyond the α phase of titanium. There is the emergence of intermetallic Ti_2Ni or Ti_4Ni_2O phases (which have the same diffraction pattern) [29, 30], and perhaps a small amount of β phase because nickel is a β-stabilizer element. It was also observed that the higher the amount of nickel was, the greater the amount of intermetallic Ti_2Ni observed in the increased intensity of the peaks and according to the system's phase diagram [23].

Cascadan et al. studied casting Ti-5Ni (wt%) [31] and Ti-10Ni (wt%) [17] concerning structural and microstructural characterization. In the case of Ti-5Ni alloy, in the XRD measurements, single α and α'phases were observed, which were corroborated by optical micrographs, showing Widmanstatten-type morphology in the samples that were subjected to quick cooling from above β transus temperature, while larger lamellar structures were observed in samples whose slow-cooling process allowed large-scale diffusion processes. In the case of Ti-10Ni alloy, the structure and microstructure of the produced alloy were analyzed by XRD and SEM, and the results showed that the alloy presents predominantly titanium α phase, with proeutectoid lamellar precipitates in eutectoid matrix of α phase and intermetallic Ti_2Ni.

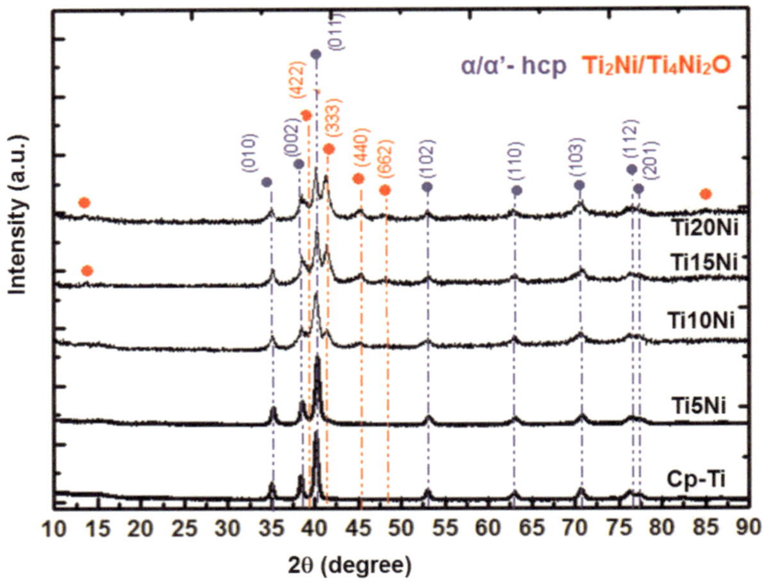

Figure 3.
X-ray diffraction for the Ti-Ni alloys after melting.

Lin et al. produced a Ti-18Ni (wt%) alloy by arc melting and showed the same phases in relation to this paper [32]. However, Ti-Ni alloys produced from metal powders melted with a 5 kW CO_2 laser presented the β phase, in addition to the α and Ti_2Ni phases, which showed that arc melting is a process of higher thermodynamic equilibrium in relation to the laser melting. The same features were observed in the case of Ti-Ni alloys that were quickly solidified for the analysis of metastable microstructures. The metastable microstructure non-equilibrium conditions also allowed the β beyond the expected α and Ti_2Ni phases [33]. In another type of processing, Ti-7Ni alloy samples were produced by sintering at 1200°C for 2 h with heating and cooling rates of 4°C/min. In this case, the peaks of X-ray diffraction of α and Ti_2Ni phases were also observed due to the low cooling rate [34]. The same occurred with the Ti-3Ni sintered to 1300°C for 2 h and heated and cooled at 4°C/min rate, with measurements of X-ray diffraction to 960°C [35]. However, samples of Ti-2Ni and Ti-5Ni sintered at 800 and 1100°C for 1 h with a heating rate of 10°C/min and cooled in the furnace presented α phase, in addition to the intermetallic Ti_2Ni and $TiNi_3$ phases. In this paper, it was found that the higher the sintering temperature and the amount of Ni are, there were higher quantities of intermetallic phases due to the diffusion process that allowed the reaction between Ti and Ni elements [36].

Peak shifts were also observed, which indicated changes in the lattice and angle parameters as well as differences in their format. The asymmetry of the lattice and angle parameters signaled distortion in the crystalline lattice because of the different quantities of substitutional and interstitial elements [15]. The α phase peaks shifted to smaller angles with the increased amount of nickel. This type of displacement is related to the increase in the lattice parameter [25] because the nickel has an atomic radius of 0.078 nm, slightly higher compared to that of titanium (0.076 nm). However, the substitutional element was not the only factor that influenced the lattice parameter; the interstitial elements and mechanical processing can also influence it [37]. In the case of the range of intermetallic phases' peaks, although the elements titanium and nickel were constant, there was a displacement of the peaks due to the presence of nitrogen and oxygen in interstitial positions.

3.2 Microstructural characterization

Figure 4 shows the optical micrographs, and **Figure 5** shows the scanning electron microscopy micrographs of all samples of Ti-Ni alloys, after melting.

For the sample with 5 wt% Ni, according to the phase diagram (**Figure 1**), there must be a certain amount of Ti_2Ni intermetallic phase, whose peak was not observed in the X-ray diffractogram because it was such a small portion, and it was not displayed by optical microscopy either due to the equipment's resolution.

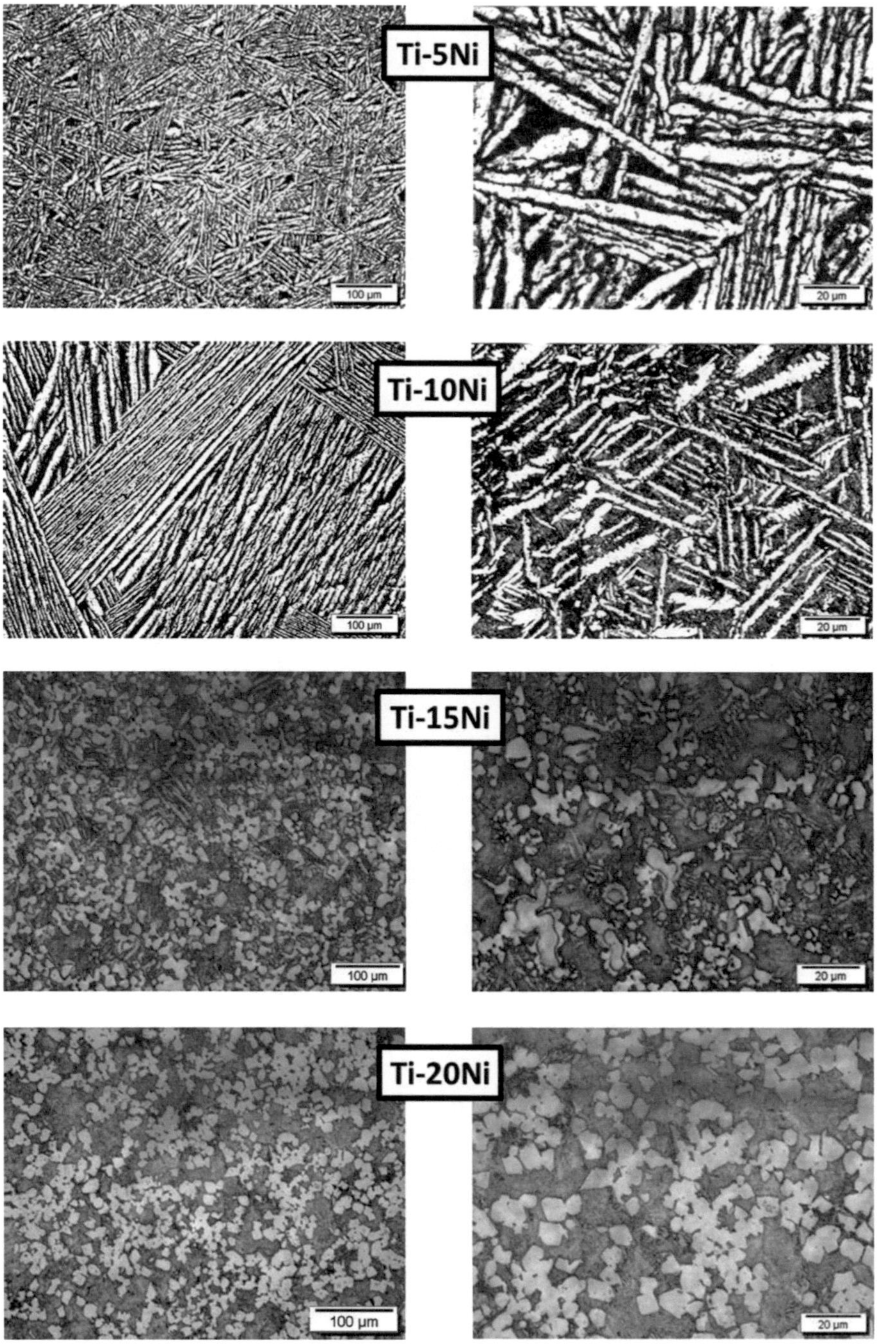

Figure 4.
Optical micrographs for Ti-Ni alloys after melting.

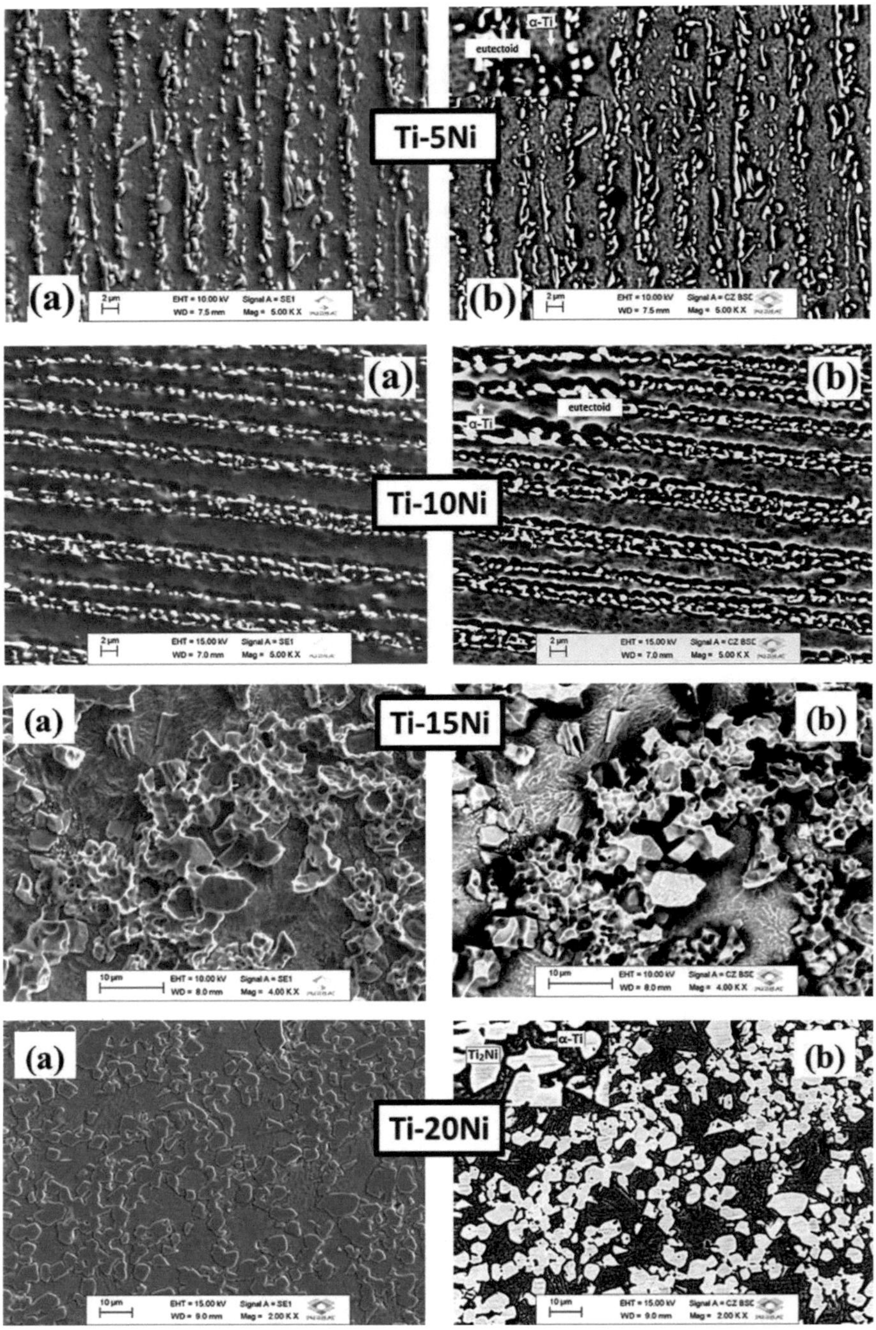

Figure 5.
SEM micrographs for the Ti-Ni alloys after melting, (a) secondary and (b) backscattering electrons.

However, the dark regions between the lamellae of the α phase clearly were a eutectoid microstructure characterized by two phases, alternating between α and Ti_2Ni itself. This eutectoid microstructure was like the microstructure of the sintered Ti-3Ni alloy shown in small expansion, but a larger magnification electron microscopy clearly showed this eutectoid microstructure [35]. In the SEM micrographs (**Figure 5**), between the lamellae of the α phase, the presence of a eutectoid microstructure characterized by two phases was observed, alternating between α and Ti_2Ni. This microstructure was considered hypoeutectoid because the concentration of 5 wt% of nickel was slightly lower than the concentration of about 6%, where the eutectoid reaction occurs [31]. The presence of the intermetallic phase that contained higher concentrations of nickel was evident in the micrographs for backscattered electrons, where the lighter area comprised the region with the highest average atomic number and the nickel had a higher atomic number compared to titanium.

For the alloy with 10 wt% of nickel, observed by optical micrographs (**Figure 4**), the microstructure consisted of precipitates of titanium proeutectoid α phase (clear) in a matrix of eutectoid microstructure (dark region) comprised of alternating lamellae of α and Ti_2Ni phases, also known as pearlite [38]. According to the phase diagram (**Figure 1**), the eutectoid reaction occurred at a nickel concentration of 6 wt%, and thus the alloy presented hypereutectoid composition, and the precipitates would be Ti_2Ni or Ti_4Ni_2O, in view of the small amount of nickel. Furthermore, α phase precipitates characterize an alloy of hypoeutectoid composition. The high quantities of oxygen and nitrogen (stabilizers of α phase) can cause change in the composition of this reaction [39]. The increase of the concentration of nickel, from the composition of about 6 wt%, does not act as a β-phase stabilizer, but causes an increase in the temperature of the formation of intermetallic phase Ti_2Ni. The images obtained by SEM (**Figure 5**) allowed more details to be obtained. At the junction of three lines where grain and leaving these needles of the α phase microstructure perpendicular to each other did originate in the grain boundaries when phase change occurred. This microstructure is similar to a Ti-24Al-11Nb alloy, whose intermetallic phase Ti_3Al locates on the grain boundaries and metallic matrix [39]. Also, according to Liu et al. [29], the diffusion of oxygen in grain boundaries is higher than in its interior, and the precipitates of Ti_4Ni_2O are formed over these. Such morphology is proven by the images obtained by backscattering electrons, where lighter points indicate the intermetallic phase between the lamellae of the α phase.

Unlike the Ti10Ni samples in the initial conditions of processing, in samples of Ti15Ni, a proeutectoid of Ti_2Ni presents precipitates which probably reacted with oxygen, thus forming Ti_4Ni_2O, comparing the amount of nickel and precipitates. This microstructure is similar to that shown by Chern lin et al. [20] and was expected due to the higher concentration of nickel in a matrix with the α phase in accordance with the X-ray diffractograms for this sample and the system's phase diagram. The optical micrographs show lighter regions that are proeutectoid of Ti_4Ni_2O precipitates in α + Ti_2Ni array. The same morphology was observed in micrographs obtained by a SEM, both for secondary electrons as backscattered, shown in **Figure 5**. Dendritic structures were also obtained by Xu et al. in Ti with 20 wt% Ni alloy procured by laser melting [24].

3.3 Mechanical characterization

Figure 6 shows the values of microhardness and elastic modulus according to the amount of nickel in the homogenized condition after hot rolling. Due to experimental conditions, the microhardness and elastic modulus measurements were made only after homogenized condition after hot-rolling conditions. There is no relationship between these properties because they involve different processes: microhardness is a measure of resistance to a plastic deformation; and elastic modulus depends on the binding force between the atoms. A factor that increases microhardness is the increase in the concentration of the substitutional element because atoms of different sizes in the crystalline lattice cause deformation in it, creating obstacles to the movement of dislocations [13]. There is also no direct

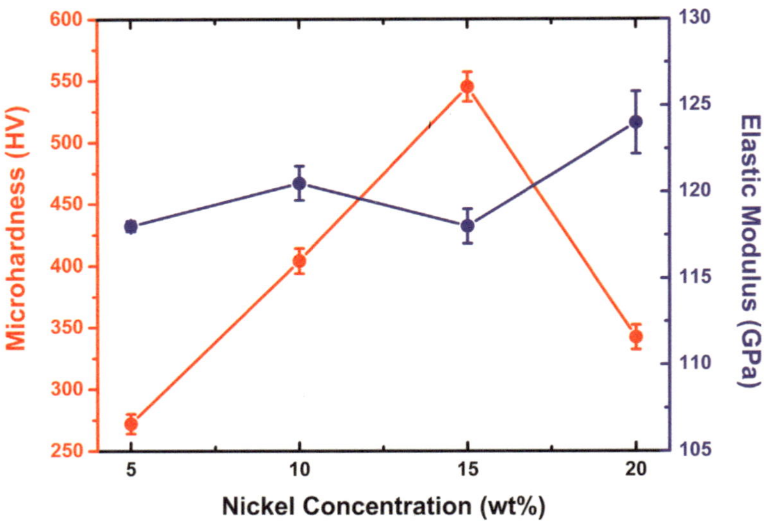

Figure 6.
Microhardness and elastic modulus values for Ti-Ni alloy samples in homogenized condition, after hot rolling.

relationship between such quantities and the amount of nickel because this is not the only factor involved.

A large variation in microhardness values was observed from 250 to 590 HV. Several factors influence a material's hardness: the concentration of substitutional and interstitial elements, microstructure, size of grain boundaries, types of phases, crystallographic orientation in which deformation occurs (since it involves a plastic deformation), and the type of processing [13]. The Ti-Ni alloy samples used in this chapter presented considerable variations of these factors.

To analyze whether these values were in agreement with those found in the literature, it was used specifically by Chern Lin et al., with similar preparation of Ti-Ni alloys by arc melting and nickel concentration ranging from 18 to 28.4 wt%. The authors obtained microhardness values ranging from 300 to 390 HV [20]. In the present study, the microhardness ranged from 345 to 390 HV, values aligned approximately with Lin's study. For Ti-Ni alloys with nickel concentration varying from 10 to 20 wt% obtained by laser fusion of metal powders [32], the microhardness ranged from 270 to 510 HV due to the increase of Ti_2Ni precipitates. Although the phases are different in relation to this study and other metallic atoms are present, the values are approximate with Lin's in relation to the other conditions, although the experimental parameters are not explicit in both cited studies.

For the Ti15Ni alloy, a high value for the microhardness was observed. As mentioned earlier, the increase of the concentration of nickel, from the composition of about 6 wt%, does not act as a β-phase stabilizer but causes an increase in the temperature of the formation of intermetallic phase Ti_2Ni that can be responsible for this increase in the hardness value [21]. In samples of Ti15Ni, a proeutectoid of Ti_2Ni presents precipitates which probably reacted with oxygen, thus forming Ti_4Ni_2O, comparing the amount of nickel and precipitates. The formation of Ti_4Ni_2O phase is another component to increase the hardness of the Ti15Ni alloy [15, 21].

The elastic modulus for Cp-Ti is around 95–105 GPa [9]. Thus, the addition of nickel caused a small increase in this property, probably from the addition of a new fcc phase referring to intermetallic Ti_2Ni. However, as shown in **Figure 6**, it was not observed as a proportional ratio of the elastic modulus with the nickel concentration.

4. Summary

The Ti-Ni samples obtained by arc melting were adequately prepared regarding stoichiometry and homogeneity and characterized by XRD, SEM, and selected mechanical properties.

The alloys showed the presence of α, Ti_2Ni, and Ti_2Ni_4O, due to the reaction of the intermetallic Ti_2Ni with oxygen. The microhardness results showed values in accordance with those presented in the literature, suitable for the use of fixation devices. There is no clear variation of the microhardness due to the amount of nickel because there were several factors involved, and the microstructures were very diversified and complex. The dynamic elastic modulus was slightly above the Cp-Ti due to the addition of a new intermetallic phase but did not vary significantly with the amount of nickel.

Acknowledgements

The authors would like to acknowledge the Brazilian agencies Capes, for D. Cascadan's fellowship, CNPq (grants #481313/2012-5 and #307.279/2013-8), and FAPESP (grant #2015/25.562-7) for their financial support.

Author details

Daniela Cascadan[1,2] and Carlos Roberto Grandini[1,2]*

1 Laboratório de Anelasticidade e Biomateriais, UNESP—Univ. Estadual Paulista, Bauru, SP, Brazil

2 IBTN/Br—Institute of Biomaterials, Tribocorrosion and Nanomedicine-Brazilian Branch, Bauru, SP, Brazil

*Address all correspondence to: carlos.r.grandini@unesp.br

References

[1] Zadpoor AA. Biomaterials and tissue biomechanics: A match made in heaven? Materials. 2017;**10**(5):528

[2] Williams D. The relationship between biomaterials and nanotechnology. Biomaterials. 2008; **29**(12):1737-1738

[3] Kolli R, Devaraj A. A review of metastable beta titanium alloys. Metals. 2018;**8**(7):506

[4] Li Y, Yang C, Zhao H, Qu S, Li X, Li Y. New developments of Ti-based alloys for biomedical applications. Materials. 2014;7(3):1709-1800

[5] Li Z, Zhao S, Ritchie RO, Meyers MA. Mechanical properties of high-entropy alloys with emphasis on face-centered cubic alloys. Progress in Materials Science. 2019;**102**:296-345

[6] Manam NS, Harun WSW, Shri DNA, Ghani SAC, Kurniawan T, Ismail MH, et al. Study of corrosion in biocompatible metals for implants: A review. Journal of Alloys and Compounds. 2017;**701**:698-715

[7] Geetha M, Singh AK, Asokamani R, Gogia AK. Ti based biomaterials, the ultimate choice for orthopaedic implants—A review. Progress in Materials Science. 2009;**54**(3):397-425

[8] Nasab MB, Hassan MR. Metallic biomaterials of knee and hip—A review. Trends in Biomaterials and Artificial Organs. 2010;**24**(1):69-82

[9] Leyens C, Peters M. Titanium and Titanium Alloys: Fundamentals and Applications. New York: Wiley-VCH; 2005

[10] Boureau G, Capron N, Tétot R. A first-principles study of dilute solutions of oxygen in titanium. Scripta Materialia. 2008;**59**(12):1255-1258

[11] Silva LMD, Claro APRA, Buzalaf MAR, Grandini CR. Influence of the substitutional solute on the mechanical properties of Ti-Nb binary alloys for biomedical use. Materials Research. 2012;**15**:355-358

[12] Noumbissi S, Scarano A, Gupta S. A literature review study on atomic ions dissolution of titanium and its alloys in implant dentistry. Materials. 2019;**12**(3): 368

[13] Callister W. Materials Science and Engineering: An Introduction. New Jersey: John Wiley & Sons; 2007

[14] Manda P, Chakkingal U, Singh AK. Hardness characteristic and shear band formation in metastable β-titanium alloys. Materials Characterization. 2014; **96**:151-157

[15] Fast JD. Gases in Metals. London: Macmillan; 1976

[16] Min X, Bai P, Emura S, Ji X, Cheng C, Jiang B, et al. Effect of oxygen content on deformation mode and corrosion behavior in β-type Ti-Mo alloy. Materials Science and Engineering A. 2017;**684**:534-541

[17] Cascadan D, Grandini CR. Influence of thermo-mechanical treatments on the structure and microstructure of alloy of titanium Ti-10%pNI obtained by arch fusion. Matéria. 2015;**20**(2): 368-373

[18] Biesiekierski A, Wang J, Abdel-Hady Gepreel M, Wen C. A new look at biomedical Ti-based shape memory alloys. Acta Biomaterialia. 2012;**8**(5): 1661-1669

[19] Sevcikova J, Pavkova Goldbergova M. Biocompatibility of NiTi alloys in the cell behaviour. Biometals. 2017;**30**(2): 163-169

[20] Chern Lin JH, Lo SJ, Ju CP. Biocorrosion study of titanium-nickel alloys. Journal of Oral Rehabilitation. 1996;**23**(2):129-134

[21] Chern Lin J-H, Moser JB, Taira M, Greener EH. Cu-Ti, Co-Ti and Ni-Ti systems: Corrosion and microhardness. Journal of Oral Rehabilitation. 1990; **17**(4):383-393

[22] Dinca VC, Soare S, Barbalat A, Dinu CZ, Moldovan A, Stoica I, et al. Nickel-titanium alloy: Cytotoxicity evaluation on microorganism culture. Applied Surface Science. 2006;**252**(13): 4619-4624

[23] Otsuka K, Ren X. Martensitic transformations in nonferrous shape memory alloys. Materials Science and Engineering A. 1999;**273-275**:89-105

[24] Lin X, Chen J, Huang W. Microstructure evolution of Ti-20wt% Ni alloy in laser solid forming. Acta Metallurgica. 2008;**44**(8):1013-1018

[25] Cullity BD, Stock SR. Elements of X-Ray Diffraction. 3rd ed. New York: Prentice Hall; 2001

[26] ASTM E384-11. Standard Test Method for Knoop and Vickers Hardness of Materials. Vol. ASTM E384-11. West Conshohocken, PA: ASTM International; 2011

[27] ASTM E92-82. Standard Test Method for Vickers Hardness of Metallic Materials. Vol. ASTM E92-82. West Conshohocken, PA: ASTM International; 2003

[28] ASTM E1876-01. Standard Test Method for Dynamic Young's Modulus, Shear Modulus, and Poisson's Ratio by Impulse Excitation of Vibration. Vol. E 1876-01. Philadelphia, USA: ASTM International; 2002

[29] Liu XP, Cao M, Jin W, Meng CG, Yang D. Effect of annealing temperature on transformation behaviors of Ti-50.2 at. pct Ni thin film. Journal of Materials Science and Technology. 2001;**17**(S1): 40-42

[30] ICSD. Inorganic Crystal Structure Database. 2014. Available from: http://icsd.ill.eu/icsd/index.php

[31] Cascadan D, Buzalaf MAR, Grandini CR. Effect of heat treatment on microstructure and mechanical properties of Ti-5wt-%Ni alloys for use as biomaterial. International Heat Treatment and Surface Engineering. 2014;**8**(3):107-110

[32] Lin X, Yue TM, Yang HO, Huang WD. Phase evolution in laser rapid forming of compositionally graded Ti-Ni alloys. Journal of Engineering Materials and Technology. 2009;**131**(4): 041002-041002-5

[33] Nagarajan R, Aoki K, Chattopadhyay K. Microstructural development in rapidly solidified Ti-Ni alloys. Materials Science and Engineering A. 1994;**179–180**(Part 1): 198-204

[34] Luo SD, Yang YF, Schaffer GB, Qian M. The effect of a small addition of boron on the sintering densification, microstructure and mechanical properties of powder metallurgy Ti-7Ni alloy. Journal of Alloys and Compounds. 2013;**555**:339-346

[35] Yang YF, Luo SD, Bettles CJ, Schaffer GB, Qian M. The effect of Si additions on the sintering and sintered microstructure and mechanical properties of Ti-3Ni alloy. Materials Science and Engineering A. 2011; **528**(24):7381-7387

[36] Panigrahi BB. Sintering behaviour of Ti-2Ni and Ti-5Ni elemental powders. Materials Letters. 2007;**61**(1):152-155

[37] Martins JRS Jr, Araújo R, Donato T, Arana-Chavez V, Buzalaf M,

Grandini C. Influence of oxygen content and microstructure on the mechanical properties and biocompatibility of Ti-15 wt%Mo alloy used for biomedical applications. Materials. 2014;7(1): 232-243

[38] Diebold TP, Aaronson HI, Franti GW. Influence of interphase boundary structure upon the mechanism of eutectoid decomposition in a Ti-Ni alloy. Metallurgical Transactions A. 1978;**9**(9):1339-1341

[39] Lütjering G, Williams JC. Titanium. 2nd ed. Berlin: Springer; 2007

Chapter 3

Cu-Based Shape Memory Alloys: Modified Structures and Their Related Properties

Safaa Najah Saud Al-Humairi

Abstract

Cu-Al-Ni shape memory alloys (SMAs) have been developed for high-temperature applications due to their ability to return to pre-deformed shape after heating above the transformation temperature, as well as these alloys have a small hysteresis and high transformation temperature comparing with other shape memory alloys. Adding some of the alloying elements such as Ti, Mn, Be, Zr and B or changing the interior content either Al or Ni by increasing/decreasing may have a significant effect on the phase transitions and enhance the mechanical properties of these alloys. However, the martensite phase transformation is the most important factor, which can be changing the whole properties of Cu-Al-Ni SMAs, where this phase is mainly affected by the alloying elements additions. This chapter reviews the effect of alloying elements on the phase transitions and the enhancement of the mechanical properties of this alloy.

Keywords: shape memory alloys, martensitic transformation, Cu-Al-Ni, grain refinement, alloying elements

1. Introduction

Shape memory alloys (SMAs) are insightful types of materials that is designed to undertake the phase transformation of martensitic phase once the thermomechanical loads are employed, and also in a position to restore their initial form as soon as heated up above particular temperatures [1–3]. A couple of symmetries take place for the structural morphologies within this kind of the materials; high symmetry at high-temperature is known as austenite and also a low symmetry at lower temperatures known as martensite. The martensitic transformation that starts for the transformation of austenite (high temperature phase) ⇄ martensite (low temperature phase), is the principal characteristic in charge of shape memory alloys features. Furthermore, this transformation is prominent as diffusionless solid-state step of transformation which is presented by means of nucleation, accompanied by the formation route of the relative austenitic phase [4, 5]. Due to the pseudo-elasticity and shape memory effect (SME) properties, the shape memory alloys have been remarkably distinguished compared to other types of materials [6–8], in which they are completely related the incidence of martensitic phase transformation. The variant crystal structure disparities take place during the phase transformation of a cubic structure (austenite) transferred to a monoclinic structure (martensite).

These sorts of martensite forms have the ability to be organized independently in a self-accommodation approach by the mechanism of twinning throughout the inadequacy of the practiced stresses, with the consequence that virtually no shape transform can certainly be realized. The results of martensitic phase deformation are able to be detwinned into a single variant corresponded to the applied loads, and consequently a large inelastic strain happened [9, 10]. Heating the deformed alloys to a certain temperature above the austenite temperature will turn the inelastic strain to be recovered through transferring the existed martensite to austenite, this kind of feature is known as shape memory effect (SME) [9]. On the other hands, the pseudo-elasticity (PE) is caused by transferring the twinned martensitic phase into detwinned phase and obtained the shape recovery under the austenite starts temperature; in other words, the deformation of loading and unloading will be occurred in the austenite phase. This kind of structure transfer will be resulted in a large inelastic strain and a consequence of the phase reverse transformation, the initial shape will be restored upon the unloading process. Therefore, these types of materials such as Ti-based, Cu-based, and Fe-based SMAs are capable to demonstrated SME and PE [11–13]. Generally, there are two groups of martensitic transformation, thermoelastic and non-thermoelastic [14]. The thermoelastic martensitic transformations happen during the mobile interfaces between the martensite phase and parent phase. These types of interfaces are able to move during the reverse martensitic transformation as an alternative to the nucleation of the parent phase, which leads to a crystallographically reversible transformation [1]. On the other hand, the non-thermoelastic martensitic transformations are mainly found in ferrous alloys, which are related to the non-mobile interfaces of the martensitic parent phase pinned by permanent defects leading to a successful nucleation and growth. As a result of the austenite re-nucleation during the reversible martensitic transformation, these kinds of transformations are crystallographically non-reversible, in which the martensite phase is not able to return to original phase [15].

2. Shape memory characteristics

2.1 Shape memory effect property

Shape memory effect (SME) is a property of SMAs which enable thermoelastic martensitic transformation. Shape memory effect will occur with the deformation of the SMA in the martensitic phase during the loading and unloading at temperatures below M_f. After heating these deformed alloys to a temperature above A_f, the austenite phase forms, and thus, the original shape is recovered. **Figure 1** shows a typical loading path $1 \rightarrow 2 \rightarrow 3 \rightarrow 4 \rightarrow 1$, wherein the property of SME is observed [16]. The parent phase transforms into the twined martensite ($1 \rightarrow 2$) when it undergoes the cooling process. The stress induced detwinning and inelastic strains can occur when the materials are loaded ($2 \rightarrow 3$). The martensite phase is in the same state of the detwinned structure without obtaining any recovered inelastic strains even after the unloaded process ($3 \rightarrow 4$). In the final step, the materials are returned to the original shape by recovering the inelastic strains after being heated above A_f ($4 \rightarrow 1$).

A self-accommodating growth of the martensitic variants ($1 \rightarrow 2$) is being produced within the stress-free cooling of austenite phase without observing any macroscopic transformation [17–19]. The essential morphology that characterize the crystallographic of these alloys is the self-accommodating structure. For instance, of the Cu-based shape memory alloys, there are 24 variants of martensite that consist of six self-accommodated groups distributed around <011> poles of

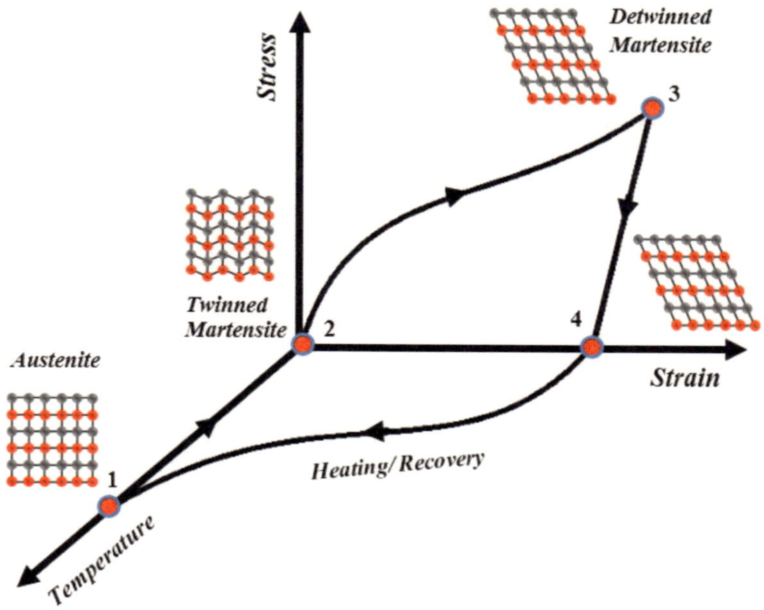

Figure 1.
Schematic diagram of stress-strain-temperature for the involved crystallographic changes during the phenomena of SME [16].

austenite which exhibit an ordinary diamond morphology. During the growth process of these groups, the macroscopic transformation strain cannot be observed, except that some of the boundaries between the martensite variants and twinning interfaces display very high movements. However, the boundary interfaces together with the detwinning structure is performed at a stress level much lower than the martensite plastic yield limit, where these phenomena is known as a reorientation of variants, which dominates at temperatures lower than M_f. In the second stage (2 → 3), the loading forces are going to reorient the variants of the martensite phase, which result in producing a large value of inelastic strain, and this strain is not recovered upon unloading (3 → 4). During the last step (4 → 1), heating the deformed alloys to a certain temperature above A_f induces reverse transformation and the inelastic strain is recovered [9, 16, 20]. The martensitic phase transformation will be unstable after the austenite finish temperature (A_f) approached without requirement for any kind of external stress. It resulted in a complete recovery will be achieved, in consequence the martensite variant reorientations do occurred, there will be an additional strain with the same value of the inelastic strain but in opposite direction, and thus, the initial shape will be recovered. Saud et al. [21] was carried out the shape memory effect test using a special designed machine, as presented in **Figure 2**; whereas the test was performed at a temperature below martensite finish temperature (i.e., 100°C), the shape recovery was obtained partially, and then it was followed by a subsequent heating above the austenite finish temperature (A_f is 300°C) using an external muffle furnace, where a full recovery was achieved.

2.2 Pseudoelasticity property

The property of pseudoelasticity in the shape memory alloys is mainly related to the induced strain recovery upon unloading at temperatures above A_f. Within the general conditions, the thermomechanical loading directions of pseudoelastic are usually started in the austenitic area at zero stress, and then move toward the region

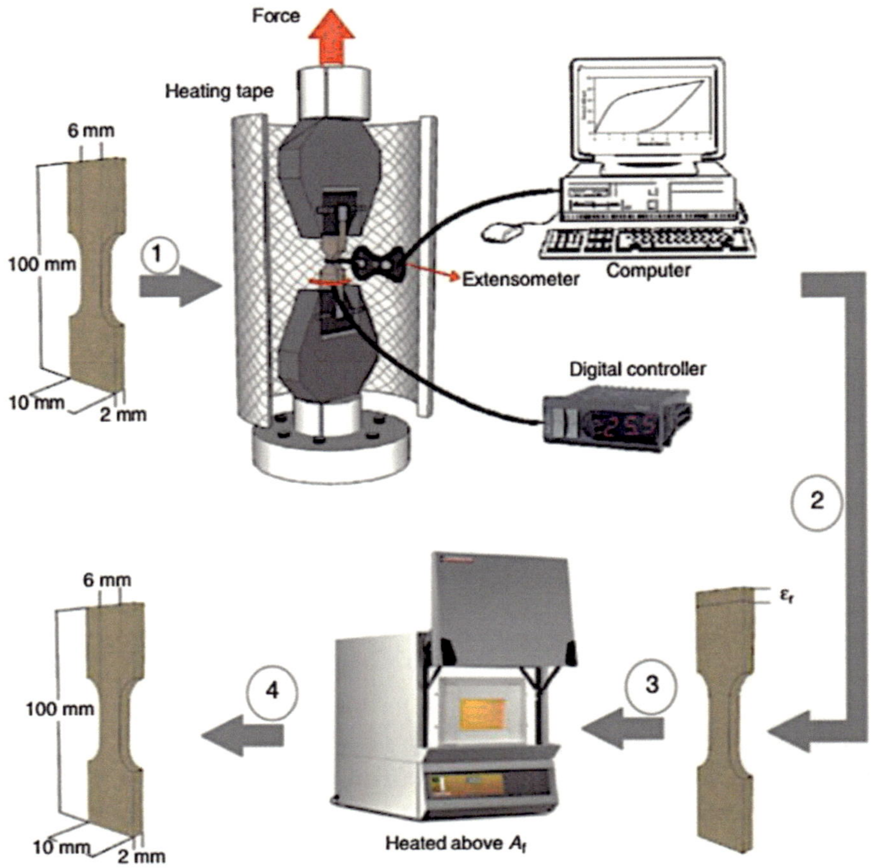

Figure 2.
Shape memory effect test [21].

of detwinned martensite, followed by the unloading toward the starting point. **Figure 3** shows the loading and unloading direction that started from point a, and moved to $b \rightarrow c \rightarrow d \rightarrow e$, then returned back to point a. Other examples are the isothermal and isobaric loading paths shown schematically in **Figure 3**.

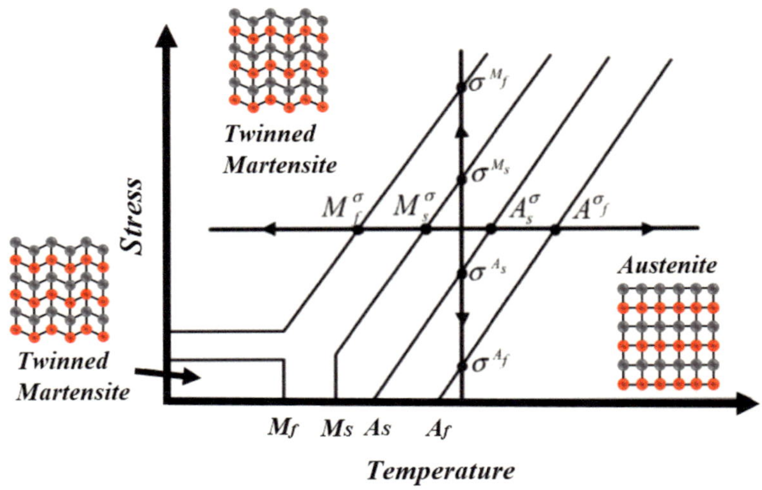

Figure 3.
The two loading paths discussed for pseudoelasticity in single crystal SMA [16].

3. Cu-based SMAs

There are two main types of Cu-based SMAs; binary alloys of Cu-Al and Cu-Zn, in which both systems performed their shape memory features in the domain of β-phase, moreover, the third element addition to the binary and/or ternary is aimed to modify and control the transformation temperatures in comprehensive range in meet the application requirements, i.e., $T \approx 100–370°C$. From this point of view, it was proven that the transformation temperatures are highly sensitive to the composition of alloys. Accuracy of 10^{-3} to 10^{-4} at.% is typically essential to obtain reproducibility more desirable than 5°C. Copper-based alloys commonly display considerably less hysteresis as compared to NiTi. Cu-Zn-Al alloy is not difficult to produce and is quite inexpensive. It decomposes into the equilibrium phases whenever overheated, therefore leading to a stabilization of the martensite. The properties of Cu-Al-Ni and Cu-Zn-Al SMAs are listed in **Table 1**. The availability of additives, including Co, Zr, B or Ti, is vital to provide grains from 50 to 100 nm in size. Add-on of boron is also used to enhance the ductility of the material. Cu-Al-Ni is substantially less vulnerable to stabilize as well as aging phenomena. This alloy performs with less hysteresis than NiTi and turns brittle as Ni increases much beyond 4 at.% [22]. It is also prevalent for Ni to be retained at a constant 4 at.% and this alloy is composed of Cu96-xAlxNi4 [23, 24]. In general, increasing the Al amount can lead to increase the stability of martensite. The purpose of the Al addition is to reduce the transformation temperatures. This variety is nearly entirely linear, ranging from M_f = 203 K and A_f = 250 K for a 14.4 at.% Al to M_f = 308 K and A_f = 348 K for a 13.6 at.% Al [22]. However, as the temperatures tend to be operated over a wide range; the sensible higher limit for transformation is 473 K. Above this temperature there is certainly an immediate degradation in the transformation as a result of aging effects. The typical Cu-based SMAs are able to exhibit a pseudoelastic strain of about of 4–6%. With the martensite to martensite transformation, very high pseudoelastic strain levels are displayed. A single crystal of the Cu81.8Al14Ni4.2 SMA can exhibit approximately 18% of the pseudoelastic strain associated with 100% of the shape recovery [25]. Cu-Zn alloy with the addition of the third element of Sn with a weight percentage of 34.7% has exhibited very low transformation temperatures, around M_f of 208 K and an A_f of 235 K [26]. As well this addition has exhibited a transformation strain (ε^t) with applied strain of 2.5% along with a pseudoelastic strain of around 8% by obtaining a full strain recovery [26]. In recent years, a minor amount (about 0.6 wt.%) of beryllium was added as a third element to the binary alloy of Cu-Al, and it was found that this addition led to reduce the transformation temperatures from 200 to 150°C with very good thermal stability.

Cu-based SMAs consist of different types of alloys, but the most frequently used alloys are Cu-Zn-Al and Cu-Al-Ni due to their inexpensive production cost and high resistance to the degradation of functional properties that occurred during the aging processes. There are many features that characterized the Cu-Al-Ni SMA rather than other shape memory alloys, such as considerably cheaper than Ni-Ti alloys and high transformation temperatures.

3.1 Phase diagram of Cu-Al-Ni SMAs

Figure 4 displayed the cross section of the ternary alloys of Cu-Al-at 3 wt.% of nickel. The alloy may possibly demonstrate shape memory characteristics as long as the martensitic transformation materialized. With the intention to ascertain undercooling, in which it vital to enforce the martensitic transformation, with a long of fully consideration that the heat treatment can never be prevented. It comes with annealing in the temperature variety of stable β phase to ensuing water quenching and resulted in the formation of β phase.

Group	No.	Alloy composition	Transformation temperature (°C)	Hysteresis (°C)	Tensile strain (%)	Strain recovery (%)	Remarks/features
Cu-based shape memory alloys	1.	Cu-Al-Ni	100–400	21.5	3–5	60–90	• Low cost • Reasonable shape memory • Good pseudoelastic behavior • Brittle in tension • Stable phase precipitation near 200°C • Reordering causes shift in transformation temperature in quenched specimen
	2.	Cu-Zn-Al	120	15–25	4	70–85	• High thermal conductivity • Reasonable recoverable shape memory strain • Inexpensive • Brittle alloys
	3.	Cu-Al-Be	150–200	20–25	3–5	80–90	• Reasonable recoverable shape memory strain • High transformation temperatures • High corrosion resistance
	4.	Cu-Al-Ni-Mn	230–280	15–20	3–4	90–100	• High shape memory behavior • Reasonable materials cost • High transformation temperatures • Good corrosion resistance
	5.	Cu-Al-Ni-Ti	120–260	12–20	2.5–4	90–100	• High shape memory behavior • Reasonable materials cost • High transformation temperatures • High corrosion resistance
	6.	Cu-Al-Ni-Fe	210–250	12–15	9	40	• Low shape memory behavior • High ductile material • Reasonable materials cost • High transformation temperatures • High corrosion resistance

Table 1.
Properties of copper-based shape memory alloys [27].

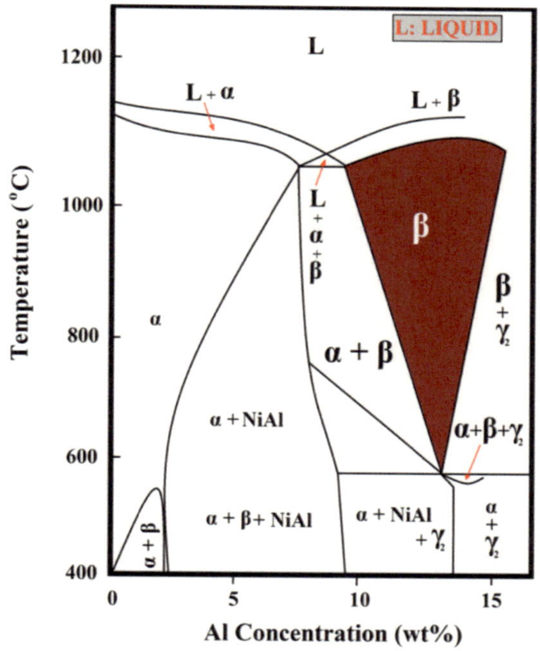

Figure 4.
Cross-section diagram of the ternary alloy of Cu-Al-3 wt.% Ni [28].

The shape memory characteristics of Cu-Al-Ni SMA are mainly dependent on the properties of the body centered cubic β phase for the binary alloys of Cu-Al [29]. During the cooling of β phase from 565°C, this phase undergoes the eutectoid decomposition of $\beta \rightarrow \alpha + \gamma_2$. However, the high cooling rates are able to prevent this phase from eutectoid decomposition and enable the martensitic transformation. When the Cu-Al-Ni SMA possess an Al content of more than 11 wt.%, the structure of body center cubic transforms to a DO3-type superlattice by transferring the β to order β_1 phase prior to martensitic transformation. In this case, the martensite "inherits" the ordered structure. At Al content between 11 and 13 wt.%, β'_1 martensite, having a monoclinic $18R_1$ structure prevails. At Al content over 13 wt.%, orthorhombic 2H-type γ'_1 martensite prevails. Which of them will appear depends on the temperature and the stress condition. In addition to these two, other types of martensite can also form (see in **Figure 5**).

The characteristic temperatures of Cu-Al-Ni alloys can lie between −200 and 200°C dependent on content of Al and Ni; the content of Al has great influence, giving them the permittivity to be used for high temperature applications. The transformation temperatures of Ni–Ti alloys can be adjusted in the range between −200 and 120°C [31]. The A_f temperature of Fe-based SMAs can increase to approximately 300°C; but at the same time, the M_s remains at room temperature or even below. The M_s temperature can be estimated using the following empirical equation [32]:

$$M_s(^{\circ}C) = 2020 - 134 \times (wt.\%Al) - 45 \times (wt.\%Ni) \quad (1)$$

The addition of Al wt.% to the Cu-based shape memory alloys can lead to reduce the transformation temperature, for instance, the addition of 14 w.% Al, the martensitic transformation start will lie around the room temperature. In spite of

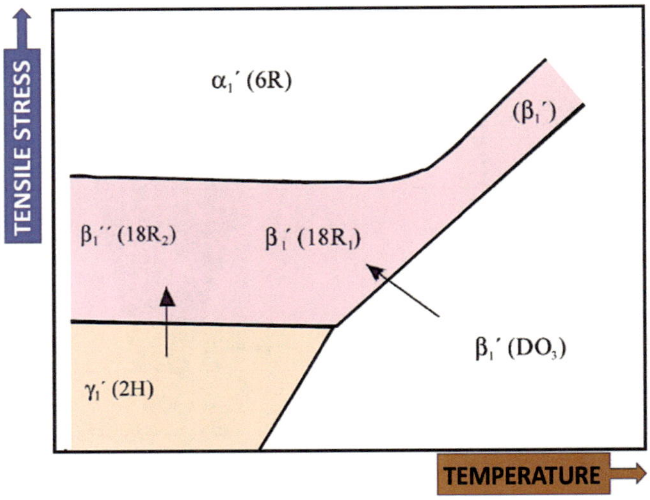

Figure 5.
Schematic phase diagram of Cu-Al-Ni alloy in temperature-stress coordinates [1, 30].

this, the Al addition may lead to from new phase known as phase γ_2 (i.e., it refers to the cubic intermetallic compound of Cu_9Al_4), in which it results in increasing the brittleness of the alloy. However, the nickel addition will play an important role of controlling the diffusion rate of Cu into Al, thereby, the may lead to retain single phase of β or β_1 till the martensitic phase transformation starts been reached during the cooling process. From another point of view, increasing the percentages of Ni in the ternary alloy of Cu-Al-Ni SMAs will be a result of the high brittleness associated with shifting the eutectoid point to higher values. Therefore, optimizing the chemical composition of the Al and Ni in the range of 14 and 3.5–4 wt.%, respectively [1]. On the other hand, these alloys still have drawbacks such as low reversible transformation that included the 4% of one-way shape memory effect and 1.5% of two-way shape memory effect. These disadvantages are mainly attributed to the intergranular cracks that occurred at a low stress level. The reasons behind the low stress failure are the large grain size, high elastic anisotropy, intense reliance of transformation strain on crystal orientations as well as segregation on grain boundaries. The first three reasons apply when there is high concentration of shear stress at the grain boundaries. The fourth reason is mainly due to weakening of grain boundaries [33].

3.2 Phase transformation morphology

The martensitic transformation can be induced both thermally and/or through applying an external stress. In other words, applying stress and decreasing the temperature both drive the austenite → martensite transformation. In fact, there is a linear relationship between the two forces that is derived from the thermodynamics relationships of the phase transformation, called the Clausius-Clapeyron relationship. Thermal treatments significantly influence the characteristics of the martensitic transformation [34], such as martensite, transformation temperatures and hysteresis, which are very sensitive to the order degree of the β phase and the precipitation process [35, 36]. The copper-based shape memory alloys exhibit a martensitic transformation from the β-phase to a close-packed structure on cooling. Additionally, the high temperatures of the β-phase for the Cu-Al-Ni alloys have a disordered bcc structure similar to the Cu-Zn-Al alloys [37]. In the Cu-Al-Ni alloys, two types of thermally induced martensites (β'_1 and γ'_1) form, depending on the alloy's composition and heat treatment [38–41]. The stability of the β-phase decreases with decreasing temperature. For example, at a lower temperature, the β-phase can remain metastable under proper cooling (air cooling) [42–44]. The stability limit of the overcooled β-phase must then be established to avoid the expansion of the ordination state of the β-phase and/or the precipitation of the stable phases. However, the improved mechanical properties of Cu-Al-Ni SMA are highly related to the production of alloys with a fine grain size [45]. During the heating-cooling processes, the structure of these alloys' changes within the martensitic region. Moreover, usable forces arise during the martensite ⇔ austenite transformation upon thermal cycling due to the shape recovery properties, which allows these alloys to be used as a component in some devices [46, 47]. The martensitic transformation requires higher energy than the reverse transformation [48].

4. Effects of alloying elements on the:

4.1 Martensitic transformation temperature of Cu-Al-Ni SMA

In copper-based shape memory alloys, the most significant factor that controls the martensite transformation is the alloy chemical composition. In commercial applications, the effect of alloying elements on the martensite transformation temperature is highly beneficial during the design of an alloy with the required characteristics [49]. Grain refiners are added to Cu-Al-Ni shape memory alloys for many reasons. These effects are both direct and indirect, such as [50] (i) the transformation temperatures are modified due to the formation of intermetallics; (ii) the remaining solid solution may increase the strength of β phase, thus leading to reduce the M_s and other temperatures; (iii) producing a chemical contribution; and (v) grain growth which occurs during annealing has an influence on the transformation temperatures.

For decreasing brittleness, one of the most important defects of Cu-Al-Ni SMAs, Itsumi et al. [51] replaced 2% of the aluminum content with Mn, which suppressed the eutectoid reaction $\beta_1 \rightarrow \alpha + \gamma_2$; Mn does not decrease the transformation temperature. At the same time, they used 1% of the Ti, which resulted in grain refinement and thus intergranular cracking can be eliminated. Karagoz and Canbay [52] studied the variations of Al and Ni percentages on the phase transformation temperatures, and have found that the forward and reverse transformation temperatures are strongly influenced by the variation of Al wt.%, therefore, higher percentage of Al exhibited lowest transformation temperatures. The variation of Ni wt.% was found to be mainly responsible for suppressing the diffusivity of Cu and Al. Chang [53] found that the M_s temperature of Cu-*x*Al-4Ni SMAs decreased significantly from 180.9 to −54.7°C when the content of Al was increased from $x = 13.0$ to 14.5 as shown in **Figure 6(a–c)**. This is consistent with the study by Recarte et al. [49], in which the M_s temperature of Cu-Al-Ni SMA depended strongly on its chemical composition, particularly with the content of Al. Cu-*x*Al-4Ni SMAs with a higher content of Al exhibiting a lower M_s temperature could be ascribed to the fact that the driving force necessary for nucleation of the γ'_1 (2H) martensite is higher than that of the β'_1 (18R) [49, 54, 55].

Sampath [50] found that addition of alloying elements and grain refiners are the main factors that can increase solid solution strengthening, as some of these elements are capable of dissolving into the solution leading to the formation of a second phase. Therefore, with the addition of a minor amount of Ti, Zr, and B to the Cu-Al-Ni SMA, the transformation temperatures are led to increase, as shown in **Figure 7(a–d)**. On the other hand, when the weight percentage of Al and Ni are

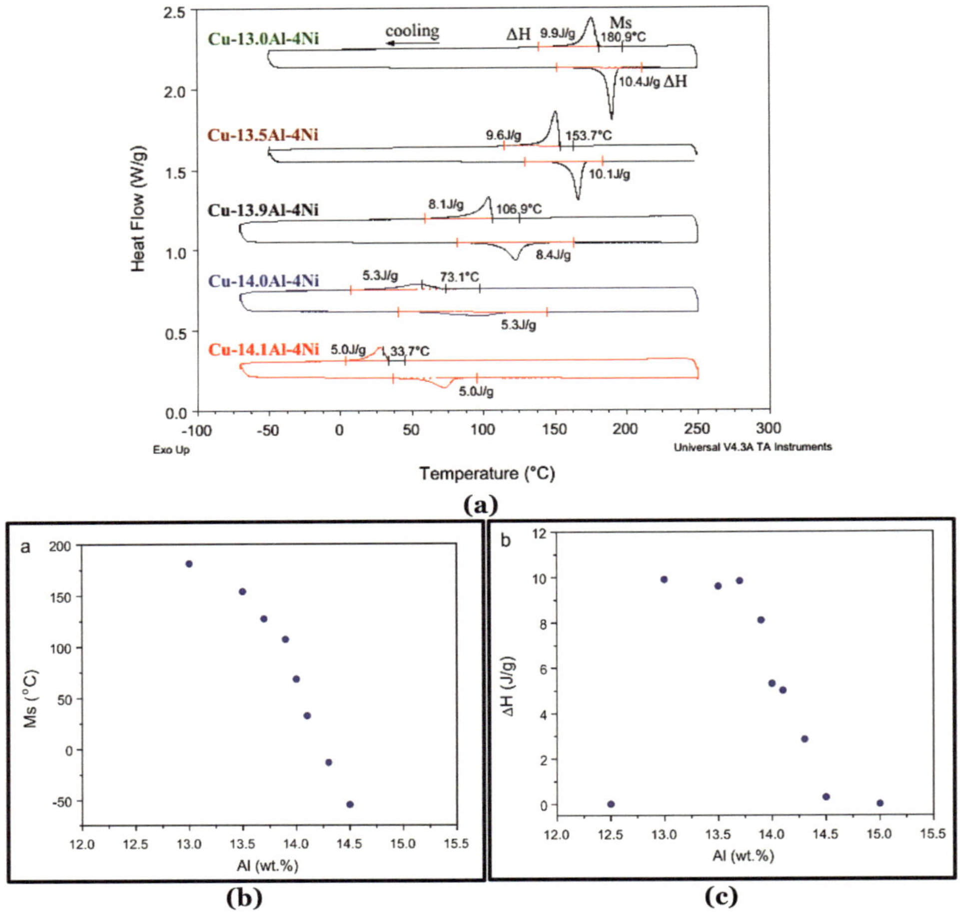

Figure 6.
Evolution of (a) DSC heating-cooling curves, (b) the M_s transformation temperature, and (c) the transformation enthalpy of the as a function of Al content [53].

decreased, the transformation temperatures increased. Thus, at less than 12 wt.% of Al, the transformation temperatures are increasing, which is in complete agreement with other researchers [23]. From the same point of view, Miyazaki et al. [56] found that with increase in the amount of Al and Ni in the entire composition of Cu-Al-Ni SMA, the transformation temperatures also tend to decrease. Sugimoto et al. [57] found that with the addition of different percentages of titanium to the Cu-Al-Ni SMA, the transformation temperature are increase. These increases are related to the presence of the X-phase as Ti-rich particles into the microstructure that can reduce the mobility of interfaces between the martensite and β phase. The martensite transformation temperature has behaved according to the type of the alloying element, where it has decreased with increasing Ti amount and increased with increasing the Zr amount as reported by Wayman and Lee [58]. This is attributed to the dissolving percentage of Ti and Zr in the β-phase. Dutkiewicz et al. [59], disagreed that Ti additions decreased the M_s. However, they have proved that the M_s temperature increases as grain size reduces, where the rapid drop of the transformation temperatures is in the smallest grain size range. Saud et al. [60] was shown that the transformation temperature of Cu-Al-Ni SMAs after the addition of Sn which was represented by the exothermic and endothermic curve in **Figure 8**, the results revealed that the behavior of the observed peak tend to be sharp and board at 232 and 350°C, respectively, due to the existence of different types of precipitates that led to limit the stability of the

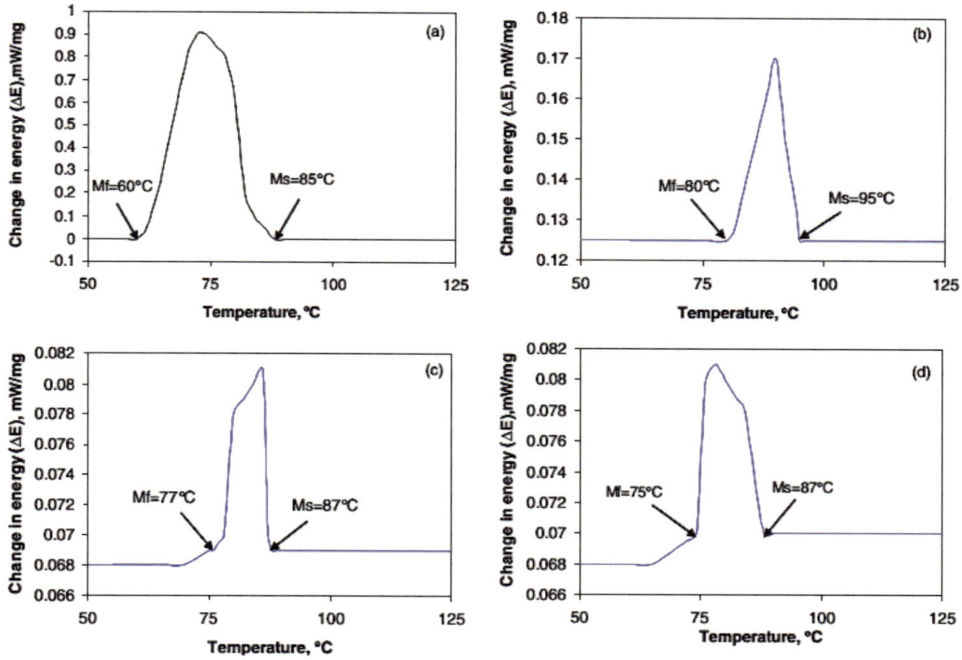

Figure 7.
Differential scanning calorimetry profiles for Cu-Al-Ni alloys: (a) Cu-Al-Ni; (b) Cu-Al-Ni-0.2Ti; (c) Cu-Al-Ni-0.4Mn; and (d) Cu-Al-Ni-0.2Zr [50].

low temperature phase and resulted in an individual transformation corresponding to the high driving force.

4.2 Martensitic structure of Cu-Al-Ni SMAs

The sort of thermally introduced martensite is totally dependent primarily on the chemical substance composition of Al and Ni in the Cu-Al-Ni SMAs. Once the martensitic transformation is produced by the deformation loading, the particular martensite acquired is determined by aspects including crystal orientation, chemical compositions of Al/Ni, deformation stress as well as applied temperature. There

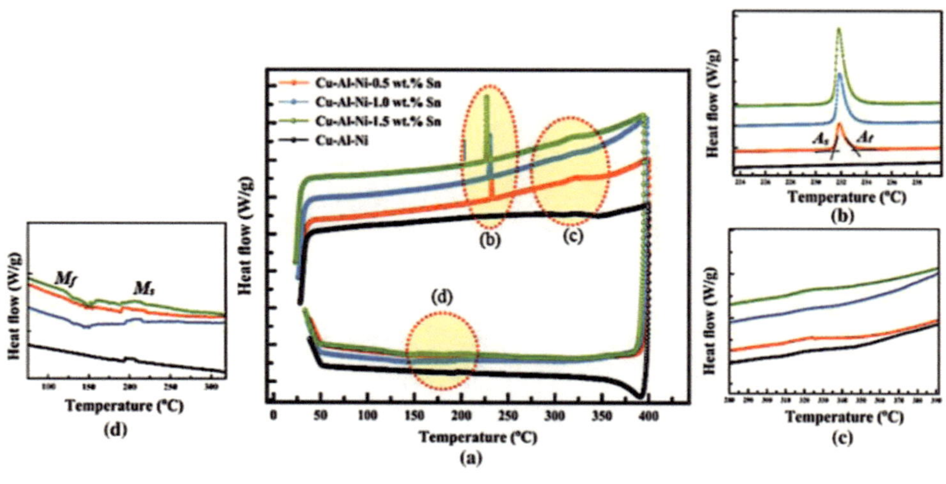

Figure 8.
Transformation temperature of Cu-Al-Ni SMAs modified with different percentage of Sn [60]; the magnified peaks of the (a,b) forwards transformation and (c) reverse transformation.

are several reasons behind adding the alloying elements to Cu-based shape memory alloys [50, 61–64], including to (1) refine the grain size, (2) restrict the martensite stabilization, (3) adjust the phase diagrams, (4) accommodate the transformation temperature, (5) improve the workability of these alloys, since they are difficult to process, due to a large grain size having formed during the solidification process, and to enhance the service life of copper shape memory alloys in applications.

The microstructure of Cu-Al-Ni SMA can be formed in a needle and/or plate-like martensites with self-accommodating morphology [50]. Two different phases are excited during adding 13.3% Al and 4.3% Ni to Cu-Al-Ni SMAs: (i) acicular morphology: β'_1; and (ii) self-accommodating morphology: γ'_1. The martensite in Cu-Al-Ni alloy has experienced a gradual transition from β'_1 to γ'_1 via a $\beta'_1 + \gamma'_1$ composition when the percentage of Al increased [49, 65]. At high cooling rate, β martensite transformed to β'_1 martensite with tiny quantities of γ'_1 phase. However, in case of low cooling rate, β'_1 transformed to γ'_1 martensite. The formation of γ'_1 martensite is inevitable irrespective of the processing conditions if the Al content is >14.2 wt.%. Minor additions to the base Cu-Al-Ni alloy tend to produce intermetallic compounds with Al, when the matrix of Al decreases resulting in the formation of β'_1 martensite. If the percentage of Al is less than 11.9 wt.%, large plates of α′ martensite will be formed. Fine plates of β'_1 martensite form when the Al content is about 11.9 wt.%. $\beta'_1 + \gamma'_1$ mixtures are observed in Cu-13.03 wt.% Al-4.09 wt.% Ni [66] and martensite formed mainly the M18R type with an orthorhombic structure [67]. However, Chentouf et al. [68] studied the microstructural and thermodynamic analysis of hypoeutectoidal Cu-Al-Ni shape memory alloys and determined that the amount of Al and Ni has a greater effect on the morphology of the precipitated phase as shown in **Figure 9**.

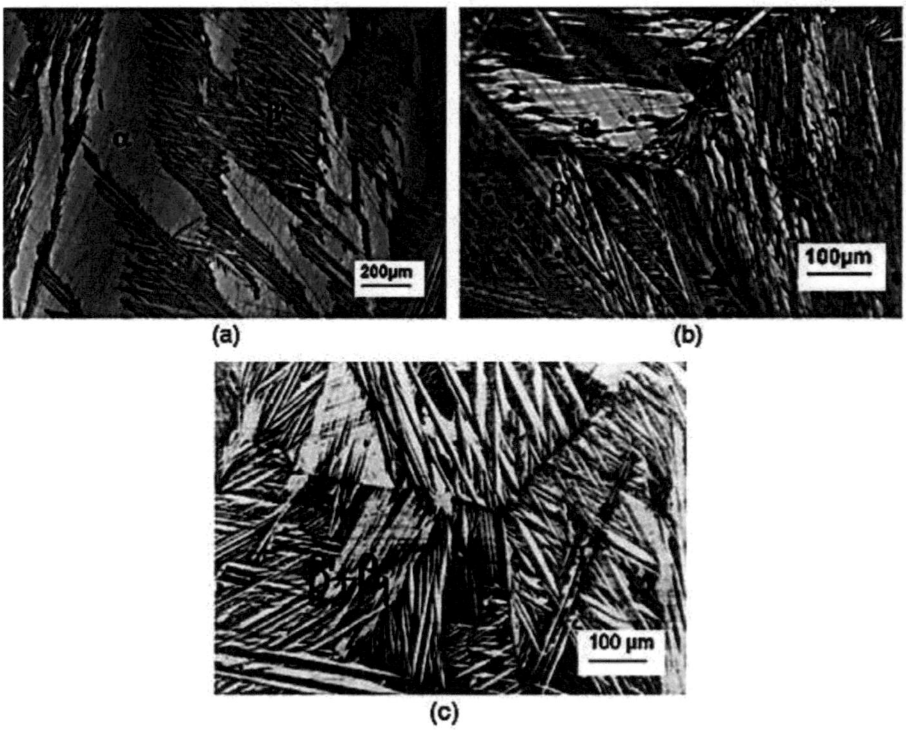

Figure 9.
Optical micrographs for alloys: (a) Cu-9.9 wt.% Al-4.43 wt.% Ni, (b) Cu-11.25 wt.% Al-4.07 wt.% Ni and (c) Cu-11.79 wt.% Al-4.37 wt.% Ni [68].

In Cu-Al-Ni shape memory alloys, large precipitate (X_L) particles are formed resulting in the transformation of the 18R basal plane order into 2H martensite at the interface of the precipitate-free and precipitate-matrix. Ratchev et al. [69] stated that there would be a change in the 18R sequence due to the modification of the stresses around the precipitates. Karagoz and Canbay [52] found that when the percentage of Al addition increased, the β phase leads to the total martensitic transformation of β'_1 and γ'_1 phases during the homogenization process and the grains formed in V-type shape along with different orientations. Chang [53] with 13 wt.% of Al, martensite exhibited self-accommodating zig zag groups at room temperature, whereas the martensite is typical β'_1 martensite with an 18R structure as shown in **Figure 10a**. However, by increasing the Al to 13.5 wt.%, a number of coarse variants of γ'_1 (2H) structure exist in the matrix of β'_1 (18R), as shown in **Figure 10b**. With further increase in the Al amount to 13.7 and 14 wt.%, the microstructure became more distinct exhibiting a β'_1 (18R) or γ'_1 (2H) martensite along with the abundant precipitate of γ_2 phase as demonstrated in **Figure 10c** and **d**. According to the relationship between the variety of transformed martensite and the composition of Cu-*x*Al-4Ni SMAs reported by Recarte [49, 54, 70], the β'_1 (18R) and the γ'_1 (2H) martensite should coexist in Cu-13.7Al-4Ni SMA, while only γ'_1 (2H) martensite exists in Cu-14.0Al-4Ni SMA.

Sugimoto et al. [57] found that with the addition of Ti to the Cu-Al-Ni SMA, a new phase known as X-phase is going to be formed which is rich in Ti-rich. Also, the volume fraction of this phase is increased linearly with increase in the percentage of Ti addition. Other work has been done by Dutkiewicz et al. [59], where they have agreed that the addition of Ti to the Cu-Al-Ni caused a smaller and elongated grain

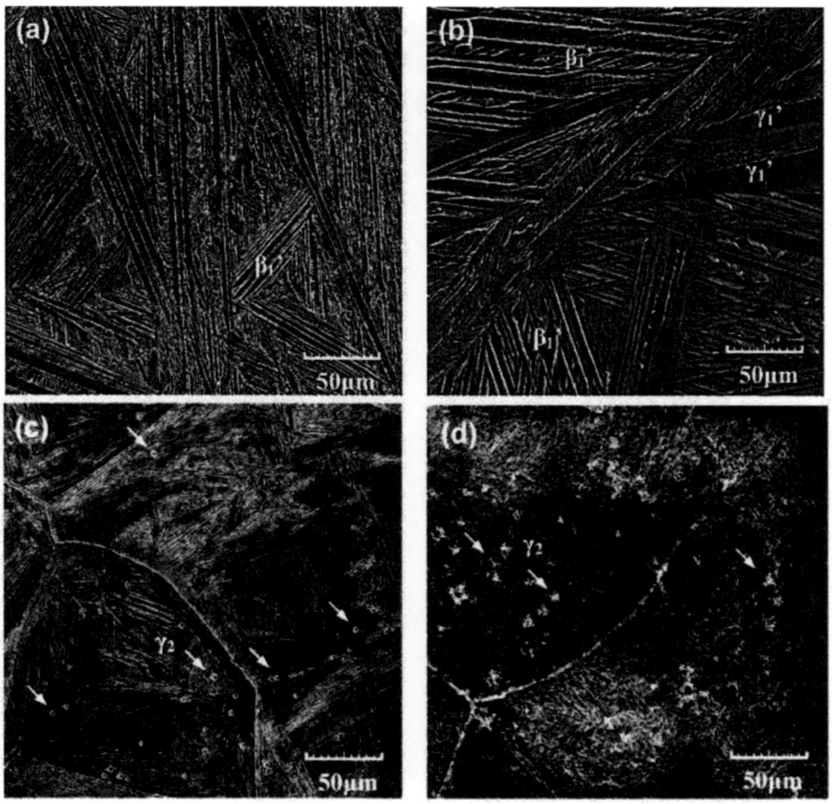

Figure 10.
SEM micrographs of (a) Cu-13.0Al-4Ni, (b) Cu-13.5Al-4Ni, (c) Cu-13.7Al-4Ni, and (d) Cu-14.0Al-4Ni SMAs [53].

size because the Ti addition restricted the grain growth as shown in **Figure 11(a–d)**. Font et al. [71] found that the addition of Mn and B along with different thermal cycling have an effect on the parameters on the martensite morphologies and orientations. They found that the martensite formed in two morphologies: plates and thin needles. The plates martensite form as self-accommodation variant groups. However, some particles have been observed to form between the plates and needles and their size is almost same with different amounts of Mn and B added. The distribution of these particles are mainly dependent on the thermal treatment conditions and by using energy dispersive spectroscopy, it was found that these particles are Mn and/or aluminum boride, a result which is in complete agreement with Morris [72]. The existence of these particles is due to difficulties dissolving Mn/B into the matrix. Sampath [50] has shown that two different morphologies are formed into the microstructure of Cu-13.3 wt.% Al-4.3 wt.% Ni SMA and these morphologies are (γ'_1 with a self-accommodating structure and β'_1 with a acicular structure). Also, it was found that with adding a minor addition of Ti, Mn, or Zr to the base alloy, new precipitations/compounds have formed with Al element as shown in **Figure 12(a–d)**. These precipitations are able to enhance the formation of martensite β'_1 phase. Saud et al. [21] presented the changes in the microstructure changes of Cu-Al-Ni SMAs after the addition of different percentages of Ti and the microstructure changes were exhibited in **Figure 13(a–d)**. it was revealed that the presence of γ'_1 and β'_1 phases, on the other hands, there is an irregular phase was observed in the modified microstructure in the shape of flower and it has been formed randomly between β'_1 plates and needles, which this phase was called as X-phase.

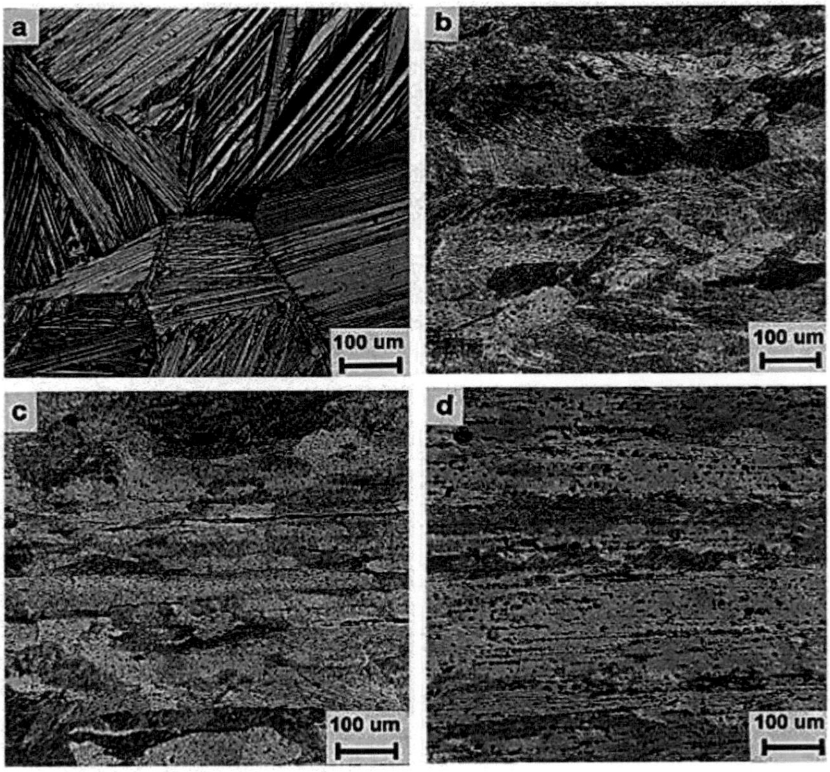

Figure 11.
Optical micrographs of (a) Cu-11.85 wt.% Al-3.2 wt.% Ni-3 wt.% Mn, (b) Cu-11.9 wt.% Al-5 wt.% Ni-2 wt.% Mn-1 wt.% Ti, (c) Cu-11.4 wt.% Al-2.5 wt.% Ni-5 wt.% Mn-0.4 wt.% Ti, and (d) Cu-11.8 wt.% Al-5 wt.% Ni-2 wt.% Mn-1 wt.% Ti [59].

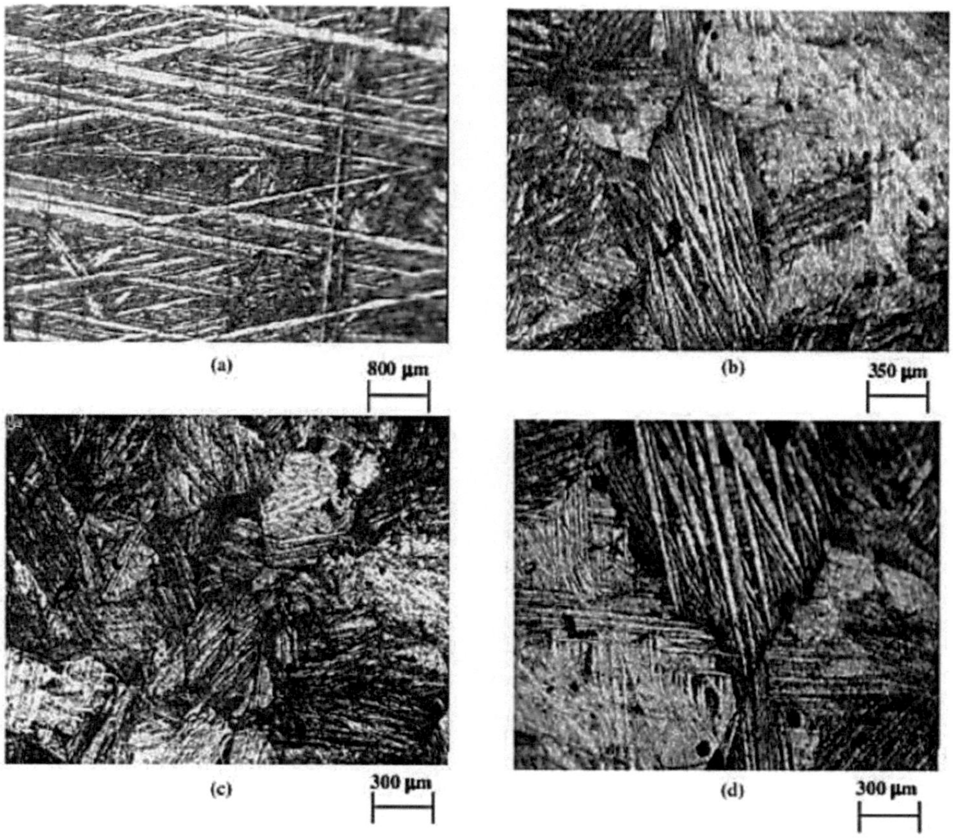

Figure 12.
Optical micrographs of Cu-Al-Ni alloys: (a) Cu-Al-Ni; (b) Cu-Al-Ni-0.2 Ti; (c) Cu-Al-Ni-0.4 Mn; and (d) Cu-Al-Ni-0.2 Zr [50].

4.3 Mechanical properties of Cu-Al-Ni SMA

Cu-Al-Ni shape memory alloys (SMA) have been selected as high potential materials for high temperature applications. This is attributed to their high thermal stability at temperatures above 100°C [73–76]. On the other hand, these alloys have their limitations such as high brittleness because of the appearance of brittle phase γ_2 at grain boundaries, the enormous increase in grain size duplicated with a high elastic variation [77–81]. Thus, their disadvantages have restricted the usage of these alloys for commercial applications [82–92]. One way to solve this problem is the grain refinement. By adding some of the alloying elements such as Ti, Mn, V, Nb, B and others or varying the compositions of Ni or Al, some improvement in mechanical properties of the conventional Cu-Al-Ni SMAs [86, 93–96] was observed. This improvement is attributed to the addition of alloying elements, where these elements are restricting the grain growth and refining the grains. However, these alloying elements have a significant effect on the mechanical properties of Cu-Al-Ni SMAs due to the formation as a second phase structure in the microstructure [97]. Miyzakai et al. [23, 56] found that varying the percentage of Al and Ni lead to changes in crack formation and propagation. It was also found that increases in the Al and Ni amount from 14 and 3.9 wt.% to 14.2 and 4 wt.% lead to the appearance of clear crack formation. This may be attributed to the amount of thermal stress induced and in accordance to the Clausius-Claperyron equation, the increase in the alloying composition of Al and Ni has an effective influence on the

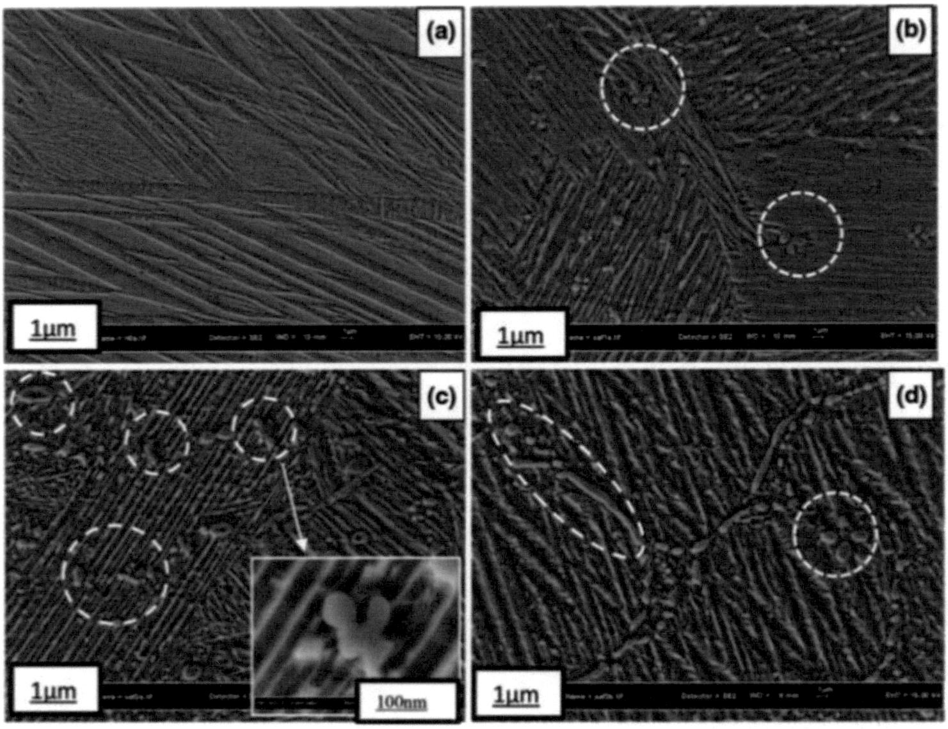

Figure 13.
FESEM micrographs showing the microstructures of the Cu-Al-Ni SMA with different concentration of Ti additions: (a) Cu-Al-Ni (alloy A), (b) Cu-Al-Ni-0.4 mass% Ti (alloy B), (c) Cu-Al-Ni-0.7 mass% Ti (alloy C), (d) Cu-Al-Ni-1 mass% Ti (alloy D) [21].

martensite thermal stress induced, which lead to crack initiation and propagation. The addition of manganese and boron efficiently refine the grain size, however, increasing of the boron concentration produced the highest strain hardening. Wayman and Lee [58] have found that the addition of boride particles helped to relieve the stress concentrations at the grain boundaries. Morris [72] found that by adding the boron to the Cu-Al-Ni SMAs, the ductility increased. This can also be attributed to the presence of boride particle. Another relevant point is that the boron addition can have an effect on the fracture mode, as it has been transferred from brittle failure to intergranular and transgranular failure. Another work by the same author [98], found that the values of yield stress, hardness and tensile

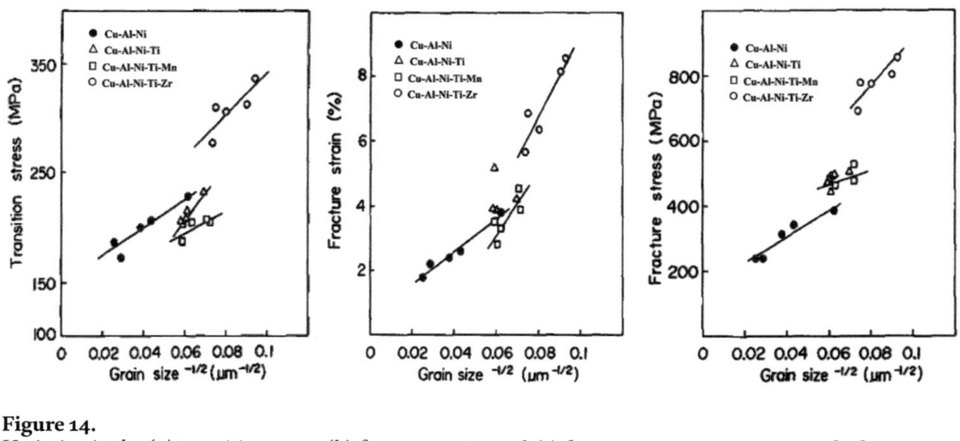

Figure 14.
Variation in the (a) transition stress, (b) fracture strain, and (c) fracture stress versus grain size [99].

strength have been increased with increasing the percentage of boron addition. It seems that the boride particles have restricted the interface movement, therefore the required stress to re-orient the martensite phase is high. These particles have played a significant role by accommodating a new strain concentration generated by the coexistence of the new stress-induced martensite. Roh et al. [99] reported that the fine grained alloys resulting from the addition of Ti, Mn, and Zr to the coarse grained Cu-Al-Ni SMA lead to enhance the fracture stress-strain. It was found that the fracture stress and strain obtained the highest value of 930 MPa and 8.6%, respectively, with the combined addition of 0.3Ti-0.6Zr to Cu-13.4AI-3.05Ni SMA. This improvement is due to grain refinement and the presence of precipitates that formed within grains in the alloy. They have also confirmed other researchers' findings [86, 100, 101] that the tensile properties of (σ_t, σ_f, and ε_t) increased as a function of decreasing grain size, as shown in **Figure 14**. In contrast, the fractured surfaces of Cu-Al-Ni SMA changed from brittle mode to different modes according to the type and amount of the addition element as shown in **Figure 15**.

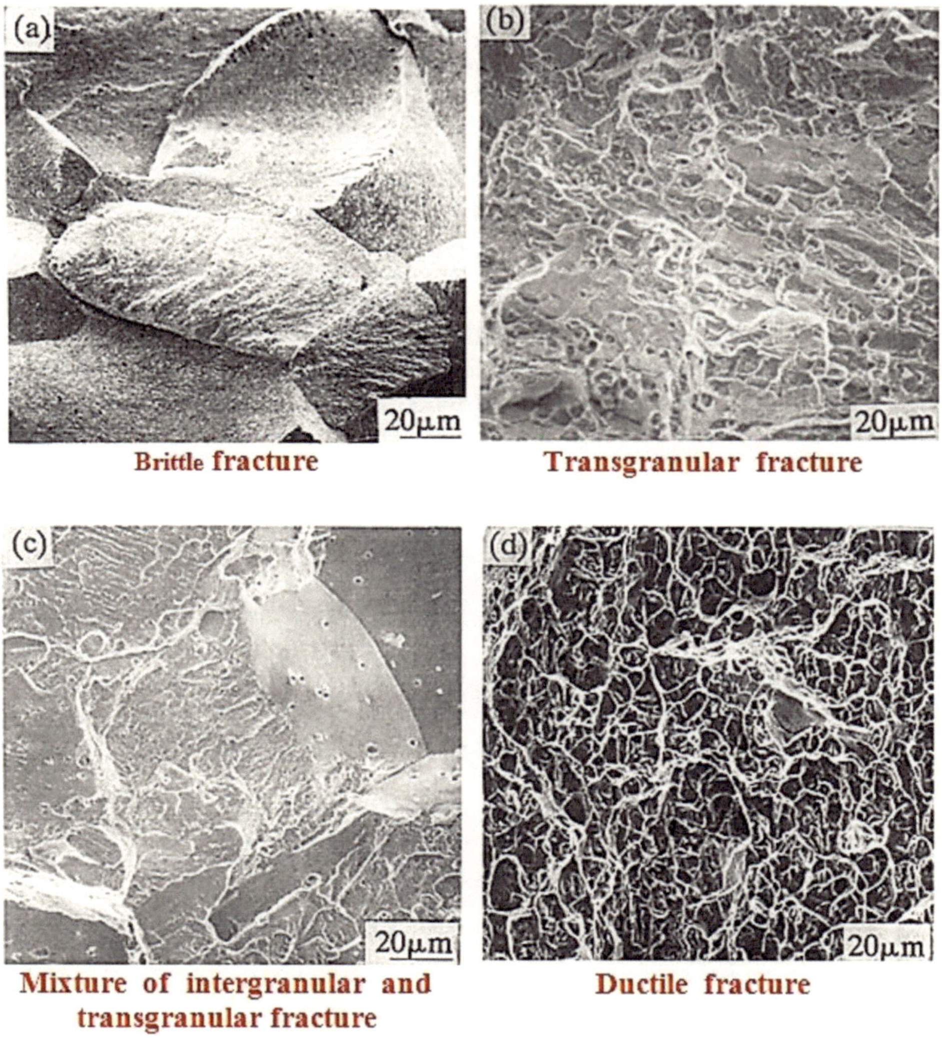

Figure 15.
Tensile fracture surfaces at room temperature for (a) Cu-13.4Al-3.8Ni SMA, (b) Cu-13.2Al-3.04Ni-0.36Ti SMA, (c) Cu-13.0Al-2.9Ni-0.36 Ti-0.22 Mn SMA and (d) Cu-13.4Al-3.05Ni-0.24Ti-0.63Zr SMA [99].

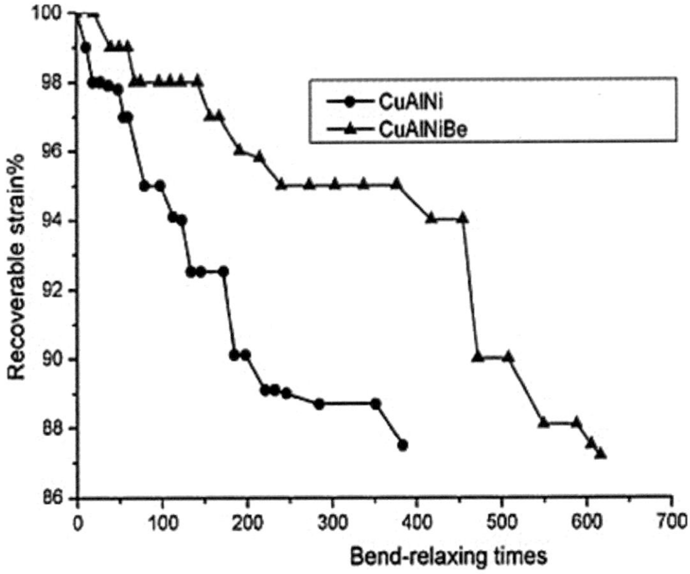

Figure 16.
Recoverable strain versus bend-relaxing time of Cu-Al-Ni and Cu-Al-Ni-Be [105].

Xu et al. [102, 103] found by adding the Be to the Cu-Al-Ni SMAs, the fatigue life has been increased, as the strain recovery has reached 30% higher than base alloy. Increase in the recovery strain is almost equal to the recovery strain of the NiTi. Zhu et al. [97] found the bending performance, tensile strength, and elongation percentage of Cu-Al-Ni-Be are higher than Cu-Al-Ni alloy, where the maximum stress of this alloy could reach to 780 MPa with 18% of strain as shown in **Figures 16** and **17**. This may imply that the mechanical property of Cu-based SMAs can be significantly improved by adding the alloying elements. The additions of Ti, Mn, and Zr to Cu-Al-Ni shape memory alloys have decreased the grain size reported by Sampath [50], therefore the values of hardness increased. This is attributed to the formation of fine precipitates that

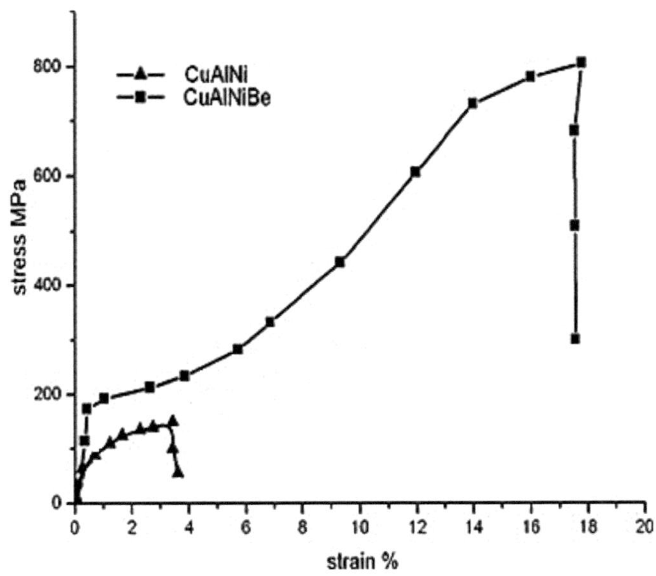

Figure 17.
Stress-strain curves of SMA samples at room temperature (25°C) [105].

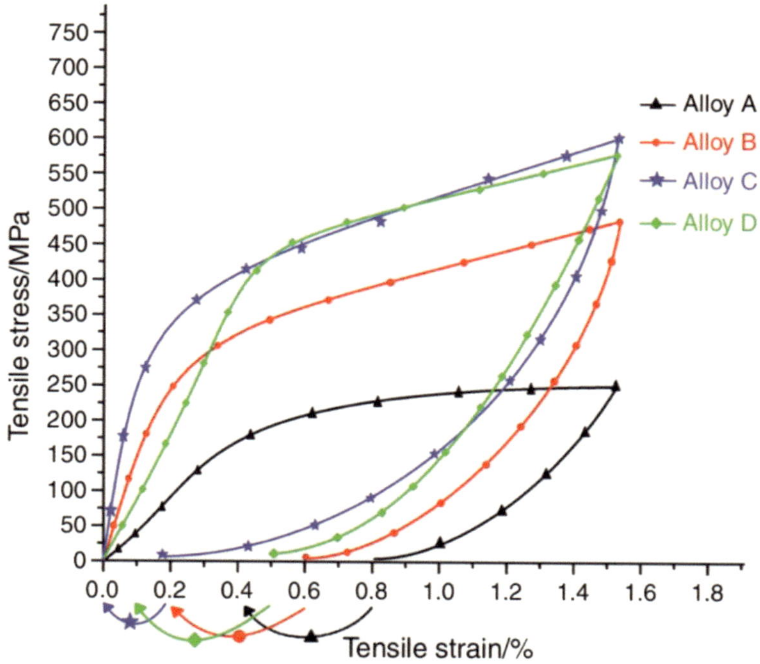

Figure 18.
Shape memory effect curves of the alloys performed at T < M_f, then preheated to T > A_f to obtain the shape recovery [21], Cu-Al-Ni (alloy A), Cu-Al-Ni-0.4 mass% Ti (alloy B), Cu-Al-Ni-0.7 mass% Ti (alloy C), Cu-Al-Ni-1 mass% Ti (alloy D).

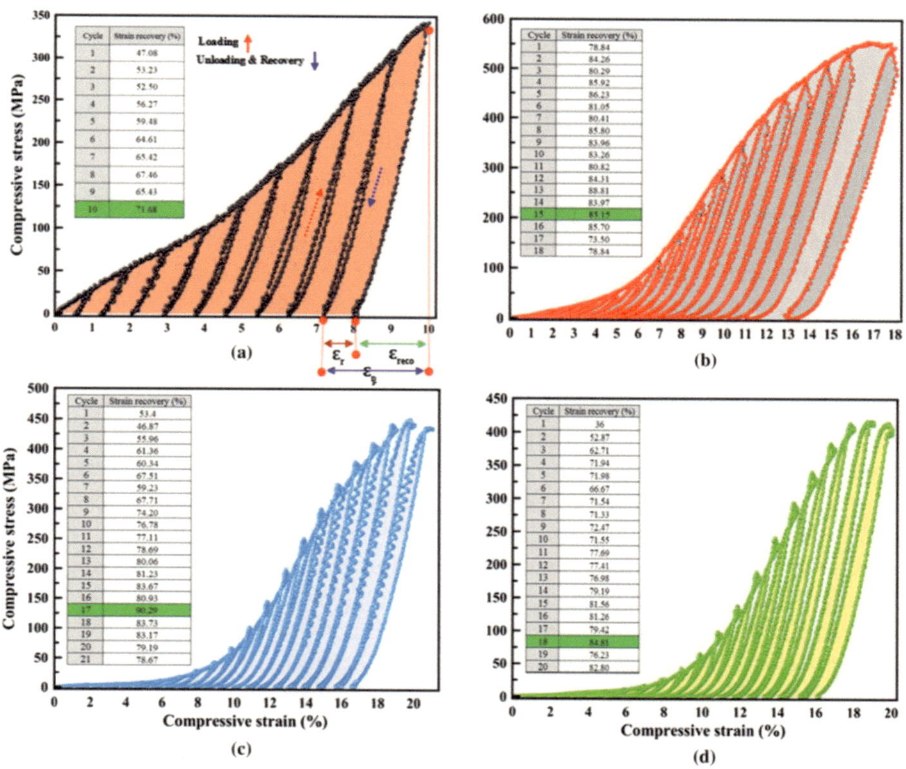

Figure 19.
Compressive stress–strain of different loading-unloading cycles tested at a temperature of 473 K (200°C); (a) Cu-Al-Ni; (b) Cu-Al-Ni-0.5 wt.% Sn; (c) Cu-Al-Ni-1.0 wt.% Sn; and (d) Cu-Al-Ni-1.5 wt.% Sn [60].

restricted the grain growth by the pinning effect. Also, other elements have shown a significant effect on the mechanical properties of Cu-Al-Ni SMAs during the addition. For example, the rupture strain of Nb and V has increased up to 14 and 6%, respectively, which is much higher than the base alloy as reported by Gomes et al. [104]. The strain recovery by the shape memory effect (ε_{SME}) of the Cu-Al-Ni SMAs with and without the Ti additions was studies by Saud et al. [21], as shown in **Figure 18**. The results were shown that the addition of Ti with different mass percentages exhibited an increase in the values of strain recovery by the SME. These enhancements in references the strain recovery were attributed to the existence of the X-phase that was brought about by the Ti additions in the parent phase. Another study by the same authors [60] shown the effect of different percentage of 0.5, 1.0, and 1.5 wt.% of Sn addition on the stress–strain curves under multi-cycles of loading and unloading. It was found that the largest number of cycles was indicated with the Cu-Al-Ni-1 wt.%Sn SMA before the occurrence of fracture, as shown in **Figure 19(a–d)**. This improvement is due to two reasons: low porosity density and the finest particle size among the alloys.

5. Brief applicability of Cu-based SMAs

Predominantly, the shape memory applications can be separated into four classes as per the essential capacity of their memory component [106–109] where the SME can be utilized to create movement as well as load, and the SE can store the twisting vitality [110, 111]. The extraordinary conduct of SMAs has produced new applications in the aviation, automobile, robotization, and control, machine, vitality, synthetic handling, warming and ventilation, security and safety, and hardware (MEMS gadgets) ventures. A part of these applications applies comparative strategies, ideas or systems, which are additionally relevant for different regions. Most of these plausible applications are secured with the economically accessible parallel Nitinol SMA, where its operational temperature run lies around inside the standard scope of ecological temperature boundaries to which a traveler vehicle might be uncovered amid administration, for example, between −40°C to approx. +125°C [112, 113]. The binary alloy system of NiTi SMA with change temperatures from −50°C to approximately to +110°C [106] performs well for various cycles inside vehicle areas in varies range of performing temperatures [113, 114], however not in areas with higher temperatures, for example, under the motor hood. The SMAs ought to have a martensite finish temperature well over the most extreme working temperatures (see the dark spotted lines in **Figure 20**) so as to work appropriately. The correlation of the change temperature scopes of the most widely recognized SMAs that the less expensive Cu-Al-Ni SMAs can play out the change with temperatures up to 200°C, however, these SMAs are fragile, unsteady, have low exhaustion quality and are not appropriate for numerous cyclic activities [106, 113, 115–119]. A wide determination of high temperature SMAs are accessible, however, these materials are known as costly for automobile applications [113].

Since the 1980_s, SMAs have been used in many different robotic systems, especially as micro-actuators or artificial muscles [123–125] as described by Furuya and Shimada [126] and Sreekumar et al. [127]. Today, most of the SMA robotic applications were biologically inspired (i.e., biomechanics) and widely utilized in biomedical areas but are also used extensively in other fields as well. The difficulties are to expand the execution and scaling down of the equipment stage and to build the insight of the coordinated framework (for example small-sizes, consistent and self-controlling). A few specialized issues were featured and should be settled, for example, clamping difficulties, miniature electrical connection (for micro robots), small strain, control issues and very low efficiency. Besides, part of these issues

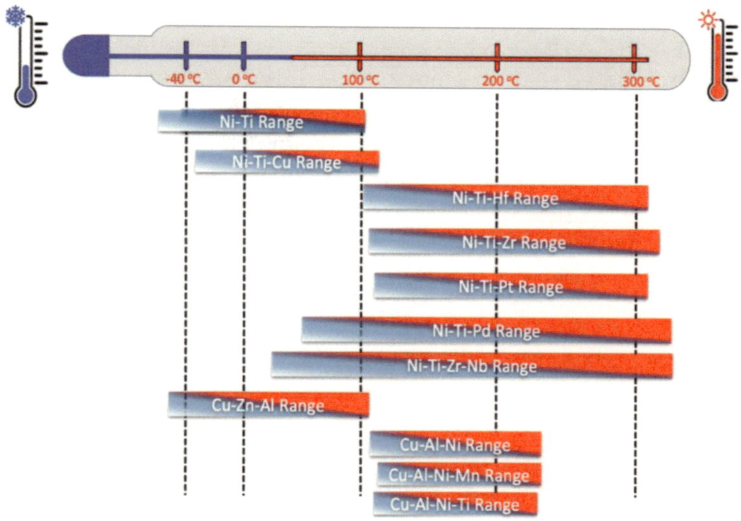

Figure 20.
Operating temperature range for automobiles applications and the transformation temperatures for selected commercially available and developed SMAs [106, 113, 120–122].

has been controlled by choosing an appropriated modeling strategy as sensor to control and feedback. For instance, the control of resistance feedback is perfect for small scale robots as it takes out the additional requirement for extra sensors, in spite of the fact that the obtained limited accuracy [127]. Earlier, it was found that the response of the SMA actuator is mainly relying on structural design and thus limited the robotic functionalities in terms of the degree of freedom (DOF). On the other hands, the heating resistivity is a common technique been implemented for a micro-size SMA actuators, however, macro-size actuators are required to a direct heating source to acquire the shape memory features. In addition to enhance the frequency of the actuators, a capacitor will be attached for a faster heating response along with above-mentioned cooling methods that resulted in larger device shape [127]. Another problem to be mentioned that the complexity of the control system that found due to the large numbers of actuator were employed to increase the robots DOF.

Another significant implementation for shape memory as absorbers in bridges that been effectively used [128, 129]. According to the study was carried out by European Union on the earthquake damages and the way of finding a sustainable technology to prevent the disaster damage via seismic vibration. A research was conducted for as modeling for four story building, which was constructed using the tendons techniques to minimize the earthquake possible damage, where it was found that the building without SMA was completely destroyed while the incorporated building with SMA got less damage, whereby, the implemented tendons were used to absorb energy based on the shape memory feature of super-elastic behavior and reduced the shocked waves of earthquake.

6. Summary

In this chapter the characteristic of shape memory alloys in terms of microstructure, mechanical properties and thermal cyclic stress-strain curves of Cu-Al-Ni SMAs. The main benefits of these alloys can be obtained after the modifications were made, as, for a certain application, the selective of shape memory alloys required the main consideration in term of manufacturing cost and performance, thus the Cu-Al-Ni SMAs have been shown interested attentions due to their low cost compared with Ti-based shape memory alloys. However, many researchers have faced a challenge when Cu-based shape memory materials are used for in many applications, due to their limitations such as the high brittleness and low recovery strain, thereby these properties need to be improved. Therefore, modifying the microstructure and the phase characteristics via adding the alloying elements may represent a more significant solution. The addition of the fourth element to the ternary alloy of Cu-Al-Ni SMA is able to alter the structure and/or morphology of the martensitic phase and thus improve the mechanical properties.

Acknowledgements

The author would like to thank the Management and Science University (MSU) for providing the research support under the Seed Research Grant No. SG-451-0518-ISE.

Conflict of interest

The authors declared without any conflict of interest.

Author details

Safaa Najah Saud Al-Humairi
Faculty of Information Sciences and Engineering, Management and Science University, Shah Alam, Selangor, Malaysia

*Address all correspondence to: safaaengineer@gmail.com

References

[1] Otsuka K, Wayman CM. Shape Memory Materials. Reprint. Illustrated ed. London, UK: Cambridge University Press; 1999

[2] Gardan J. Smart materials in additive manufacturing: State of the art and trends. Virtual and Physical Prototyping. 2019;**14**(1):1-18

[3] Farag SG. Application of smart structural system for smart sustainable cities. In: 4th MEC International Conference on Big Data and Smart City (ICBDSC). USA: IEEE; 2019

[4] Morales-Rivas L et al. Crystallographic examination of the interaction between texture evolution, mechanically induced martensitic transformation and twinning in nanostructured bainite. Journal of Alloys and Compounds. 2018;**752**:505-519

[5] Wu S et al. Mechanical properties of low-transformation-temperature weld metals after low-temperature postweld heat treatment. Science and Technology of Welding and Joining. 2019;**24**(2):112-120

[6] Vokoun D, Kafka V. Mesomechanical modeling of shape memory effect. In: Symposium on Smart Structures and Materials. California, USA: International Society for Optics and Photonics; 1999

[7] Malinin V et al. Development of methods of structural-analytical mesomechanics that take into account the statistical properties of martensitic transformations in materials with shape memory effect. In: IOP Conference Series: Materials Science and Engineering. Bristol, England: IOP Publishing; 2018

[8] Yu C et al. A micromechanical model for the grain size dependent super-elasticity degeneration of NiTi shape memory alloys. Mechanics of Materials. 2018;**125**:35-51

[9] Kumar P, Lagoudas D. Introduction to Shape Memory Alloys. United States: Springer; 2008

[10] Mohammed MT, GEETHA M. Effect of thermo-mechanical processing on microstructure and electrochemical behavior of Ti–Nb–Zr–V new metastable β titanium biomedical alloy. Transactions of Nonferrous Metals Society of China. 2015;**25**(3):759-769

[11] Nishiyama Z, Fine ME, Wayman CM. Martensitic Transformation. United States: Academic Press; 1978

[12] Mori M et al. Tuning strain-induced γ-to-ε martensitic transformation of biomedical Co–Cr–Mo alloys by introducing parent phase lattice defects. Journal of the Mechanical Behavior of Biomedical Materials. 2019;**90**:523-529

[13] Wang J et al. New insights on nucleation and transformation process in temperature-induced martensitic transformation. Materials Characterization. 2019;**151**:267-272

[14] Popov PA. Constitutive Modelling of Shape Memory Alloys and Upscaling of Deformable Porous Media. United States: Texas A&M University; 2005

[15] Patoor E et al. Shape memory alloys, part I: General properties and modeling of single crystals. Mechanics of Materials. 2006;**38**(5):391-429

[16] Lagoudas DC. Shape Memory Alloys: Modeling and Engineering Applications. United States: Springer; 2008

[17] Saburi T et al. The shape memory mechanism in 18R martensitic alloys. Acta Metallurgica. 1980;**28**(1):15-32

[18] Alkan S et al. Transformation stress of shape memory alloy CuZnAl: Non-Schmid behavior. Acta Materialia. 2018;**149**:220-234

[19] Carpinteri A et al. Mechanical behaviour and phase transition mechanisms of a shape memory alloy by means of a novel analytical model. Acta Mechanica et Automatica. 2018;**12**(2):105-108

[20] Lagoudas DC. Shape Memory Alloys and Engineering Applications. New York: Springer; 2008

[21] Saud S et al. Influence of Ti additions on the martensitic phase transformation and mechanical properties of Cu–Al–Ni shape memory alloys. Journal of Thermal Analysis and Calorimetry. 2014;**118**(1):111-122

[22] Tadaki T. Cu-based shape memory alloys. Shape memory materials. United Kingdom: Cambridge University Press; 1998. p. 97-116

[23] Miyazaki S, Kawai T, Otsuka K. Study of fracture in Cu-Al-Ni shape memory bicrystals. Le Journal de Physique Colloques. 1982;**43**:C4, C4-813-C4-818

[24] Horikawa H et al. Orientation dependence of β1 → β1′ stress-induced martensitic transformation in a Cu-AI-Ni alloy. Metallurgical Transactions A. 1988;**19**(4):915-923

[25] Otsuka K, Sakamoto H, Shimizu K. Successive stress-induced martensitic transformations and associated transformation pseudoelasticity in Cu-Al-Ni alloys. Acta Metallurgica. 1979;**27**(4):585-601

[26] Eisenwasser J, Brown L. Pseudoelasticity and the strain-memory effect in Cu-Zn-Sn alloys. Metallurgical Transactions. 1972;**3**(6):1359-1363

[27] Lexcellent C. Shape-Memory Alloys Handbook. New Jersey, United States: Wiley; 2013

[28] Dunne NFKADP. Shape strains associated with thermally-induced and stress-induced martensite in a Cu-Al-Ni shape memory alloy. Acta Metallurgica. 1982;**30**:429-435

[29] Chen Y et al. Shape memory and superelasticity in polycrystalline Cu–Al–Ni microwires. Applied Physics Letters. 2009;**95**(17):171906

[30] Delaey L. Diffusionless transformations. In: Cahn RW, Haasen P, Kramen EJ, editors. Phase Transformations in Materials, Material Science and Technologies. New York: VCH Publishers; 1990. pp. 339-404

[31] Mavroidis C, Pfeiffer C, Mosley M. 5.1 conventional actuators, shape memory alloys, and electrorheological fluids. Automation, Miniature Robotics, and Sensors for Nondestructive Testing and Evaluation. 2000;**4**:189

[32] Scherngell H. Stability and Optimization of the Two-Way Effect in Ni–Ti and Cu–Al–Ni Shape Memory Alloys. Montanuniversit: Leoben; 2000

[33] Miyazaki S, I.C.f.M. Sciences. In: Fremond M, Miyazaki S, editors. Shape Memory Alloys. United States: Springer; 1996

[34] Zhang Xiangyang SQ, Shouwen Y. A non-invariant plane model for the interface in CuAlNi single crystal shape memory alloys. Journal of the Mechanics and Physics of Solids. 2000;**48**:2163-2182

[35] Nó ML, Caillard D, San Juan J. A TEM study of martensite habit planes and orientation relationships in Cu–Al–Ni shape memory alloys using a fast Δg-based method. Acta Materialia. 2009;**57**(4):1004-1014

[36] Qiao L et al. Nonlocal superelastic model of size-dependent hardening and dissipation in single crystal Cu-Al-Ni shape memory alloys. Physical Review Letters. 2011;**106**(8):085504

[37] Xu JW. Effects of Gd addition on microstructure and shape memory effect of Cu–Zn–Al alloy. Journal of Alloys and Compounds. 2008;**448**(1-2):331-335

[38] MLaRR C. Isothermal decomposition of some β Cu-Zn-Al alloys with e/a=1.48. Materials Science and Engineering A. 1999;**273-275**:577-580

[39] Agrawal A, Dube RK. Methods of fabricating Cu-Al-Ni shape memory alloys. Journal of Alloys and Compounds. 2018;**750**:235-247

[40] Saud SN et al. Effects of quenching media on phase transformation characteristics and hardness of Cu-Al-Ni-Co shape memory Alloys. Journal of Materials Engineering and Performance. 2015;**24**(4):1522-1530

[41] Gustmann T et al. Properties of Cu-based shape-memory alloys prepared by selective laser melting. Shape Memory and Superelasticity. 2017;**3**(1):24-36

[42] Gera DB et al. The influence of sintering parameters in the microstructure and mechanical properties of a Cu–Al–Ni–Mn–Zr shape memory alloy. Advanced Engineering Materials. 2018;**20**(10):1800372

[43] Velmurugan C, Senthilkumar V. The effect of Cu addition on the morphological, structural and mechanical characteristics of nanocrystalline NiTi shape memory alloys. Journal of Alloys and Compounds. 2018;**767**:944-954

[44] Li D-Y et al. Superelasticity of Cu–Ni–Al shape-memory fibers prepared by melt extraction technique. International Journal of Minerals, Metallurgy, and Materials. 2016;**23**(8):928-933

[45] Otsuka K, Saxena A, Deng J, Ren X. Mechanism of the shape memory effect in martensitic alloys: An assessment. Philosophical Magazine. 2011;**91**(36):4514-4535

[46] Bayram Ü, Maraşlı N. Thermal conductivity and electrical resistivity dependences on growth rate in the directionally solidified Al–Cu–Ni eutectic alloy. Journal of Alloys and Compounds. 2018;**753**:695-702

[47] Braga FO et al. Martensitic transformation under compression of a plasma processed polycrystalline shape memory CuAlNi alloy. Materials Research. 2017;**20**(6):1579-1592

[48] Wei ZG, Peng HY, Yang DZ, Chung CY, Lai JKL. Reverse transformations in CuA1NiMnTi alloy at elevated temperatures. Acta Materialia. 1996;**44**(3):1189-1199

[49] Recarte V et al. Dependence of the martensitic transformation characteristics on concentration in Cu–Al–Ni shape memory alloys. Materials Science and Engineering: A. 1999;**273**:380-384

[50] Sampath V. Studies on the effect of grain refinement and thermal processing on shape memory characteristics of Cu–Al–Ni alloys. Smart Materials and Structures. 2005;**14**(5):S253-S260

[51] Itsumi Y, Miyamoto Y, Takashima T, Kamei K, Sugimoto K. The effects of ageing on the martensitic transformation temperature in Cu-Al-Ni-Mn-Ti shape memory alloys. Advanced Materials Research. 1991;**56-58**:469-474

[52] Karagoz Z, Canbay CA. Relationship between transformation temperatures and alloying elements in Cu–Al–Ni shape memory alloys. Journal of Thermal Analysis and Calorimetry. 2013;**114**(3):1069-1074

[53] Chang SH. Influence of chemical composition on the damping characteristics of Cu–Al–Ni shape memory alloys. Materials Chemistry and Physics. 2011;**125**(3):358-363

[54] Recarte V et al. Study by resonant ultrasound spectroscopy of the elastic constants of the β phase in Cu-Al-Ni shape memory alloys. Materials Science and Engineering: A. 2004;**370**(1-2):488-491

[55] Recarte V et al. Vibrational and magnetic contributions to the entropy change associated with the martensitic transformation of Ni-Fe-Ga ferromagnetic shape memory alloys. Journal of Physics. Condensed Matter. 2010;**22**(41):416001

[56] Miyazaki S et al. The fracture of Cu–Al–Ni shape memory alloy. Transactions of the Japan Institute of Metals. 1981;**22**(4):244-252

[57] Sugimoto K et al. Grain-refinement and the related phenomena in quaternary Cu-Al-Ni-Ti shape memory alloys. Le Journal de Physique Colloques. 1982;**43**(C4):C4-761-C4-766

[58] Wayman CM, Lee JS. Grain refinement of a Cu-Al-Ni shape memory alloys by Ti and Zr additions. Transactions of Japan Institute of Metals. 1986;**27**(8):584-591

[59] Dutkiewicz J, Czeppe T, Morgiel J. Effect of titanium on structure and martensic transformation in rapidly solidified Cu–Al–Ni–Mn–Ti alloys. Materials Science and Engineering: A. 1999;**273**:703-707

[60] Saud SN et al. Influence of tin additions on the phase-transformation characteristics of mechanical alloyed Cu-Al-Ni shape-memory alloy. Metallurgical and Materials Transactions A. 2016;**47**(10):5242-5255

[61] van Humbeeck J, Chandrasekaran M, Stalmans R. Copper-based shape memory alloys and the martensitic transformation. Proceedings of the International conference on Martensite Transformation. 1993;**25**:1015

[62] van Humbeeck J, Chandrasekaran M, Stalmans R. Shape memory alloys, types and functionalities. In: Encyclopedia of Smart Materials. United States: John Wiley & Sons, Inc; 2002

[63] Chen X et al. Microstructure, superelasticity and shape memory effect by stress-induced martensite stabilization in Cu–Al–Mn–Ti shape memory alloys. Materials Science and Engineering: B. 2018;**236**:10-17

[64] Guzik AT, Benafan O. Design and Development of CubeSat Solar Array Deployment Mechanisms Using Shape Memory Alloys. In: 44th Aerospace Mechanisms Symposium. Ohio: NASA/TM; 16-18 May 2018. pp.219914

[65] Bhattacharya B, Bhuniya A, Banerjee MK. Influence of minor additions on characteristics of Cu–Al–Ni alloy. Materials Science and Technology. 1993;**9**:654-665

[66] Chen C, Liu T. Phase transformations in a Cu-14.2 Al-7.8 Ni alloy. Metallurgical and Materials Transactions A. 2003;**34**(3):503-509

[67] Duerig TW et al. Engineering Aspects of Shape Memory Alloys. United Kingdom: Elsevier; 1990

[68] Chentouf SM et al. Microstructural and thermodynamic study of hypoeutectoidal Cu–Al–Ni shape memory alloys. Journal of Alloys and Compounds. 2009;**470**(1-2):507-514

[69] Ratchev P, Van Humbeeck J, Delaey L. On the formation of 2H stacking sequence in 18R martensite plates in a precipitate containing CuAlNiTiMn alloy. Acta Metallurgica et Materialia. 1993;**41**(**8**):2441-2449

[70] Recarte V, Hurtado I, Herreros J, Nó ML, San Juan J. Precipitation of the stable phases in Cu-Al-Ni shape memory alloys. Scripta Materialia. 1996;**34**:255-260

[71] Font J et al. Thermal cycling effects in high temperature Cu-Al-Ni-Mn-B shape memory alloys. Journal of Materials Research. 1997;**12**(9):2288-2297

[72] Morris MA. Microstructural influence on ductility and shape memory effect of some modified Cu.Ni.Ai alloys. Scripta Metallugica. 1991;**25**:1409-1414

[73] Lojen G et al. Microstructure of rapidly solidified Cu–Al–Ni shape memory alloy ribbons. Journal of Materials Processing Technology. 2005;**162-163**:220-229

[74] Fremond M, Miyazaki S. Shape Memory Alloys. Wien, New York: Springer-Verlag; 1996

[75] Prasad K et al. Metallic biomaterials: Current challenges and opportunities. Materials. 2017;**10**(8):884

[76] Formentini M, Lenci S. An innovative building envelope (kinetic façade) with shape memory Alloys used as actuators and sensors. Automation in Construction. 2018;**85**:220-231

[77] Otsuka K, Wayman CM. Shape Memory Materials. Cambridge: Cambridge University Press; 1998

[78] Miyazaki KOS, Sakamoto H, Shimizu K. Study of fracture in Cu-Al-Ni shape memory bicrystals. Transactions of the Japan Institute of Metals. 1981;**22**:244-252

[79] Husain SW, Clapp PC. Grain boundary embrittlement in Cu-AI-Ni β phase alloys. Journal of Materials Science. 1987;**22**:2351-2356

[80] Dalvand P et al. Properties of rare earth added Cu–12wt% Al–3wt% Ni–0.6 wt% Ti high temperature shape memory alloy. Materials Science and Engineering: A. 2019;**754**:370-381

[81] Edwards TEJ et al. Slip bands in lamellar TiAl during high cycle fatigue microcompression by correlative total strain mapping, diffraction orientation mapping and transmission electron imaging. International Journal of Fatigue. 2019;**124**:520-527

[82] Lee JS, Wayman CM. Grain refinement of Cu-Zn-Al shape memory alloys. Metallography. 1986;**19**(4):401-419

[83] Sure GN, Brown LC. The mechanical properties of grain refined β-Cu-Al-Ni strain-memory alloys. Metallurgical and Materials Transactions A. 1984;**15**:1613-1621

[84] Bhattacharya S, Bhuniya A, Banerjee MK. Influence of minor additions on characteristics of Cu-Al-Ni alloy. Materials Science and Technology. 1993;**9**(8):654-658

[85] Adachi K, Hamada Y, Tagawa Y. Crystal structure of the X-phase in grain-refined Cu-Al-Ni-Ti shape memory alloys. Scripta Metallurgica. 1987;**21**(4):453-458

[86] Kim J et al. Effects on microstructure and tensile properties of a zirconium addition to a Cu-Al-Ni shape memory alloy. Metallurgical and Materials Transactions A. 1990;**21**(2):741-744

[87] Morris MA, Lipe T. Microstructural influence of Mn additions on thermoelastic and pseudoelastic properties of Cu-Al-Ni alloys. Acta Metallurgica et Materialia. 1994;**42**(5):1583-1594

[88] Gao Y, Zhu M, Lai JKL. Microstructure characterization and effect of thermal cycling and ageing on vanadium-doped Cu–Al–Ni–Mn

high-temperature shape memory alloy. Journal of Materials Science. 1998;**33**(14):3579-3584

[89] Oliveira J et al. Microstructure and mechanical properties of gas tungsten arc welded Cu-Al-Mn shape memory alloy rods. Journal of Materials Processing Technology. 2019;**217**:93-100

[90] Yang S et al. Excellent superelasticity and fatigue resistance of Cu-Al-Mn-W shape memory single crystal obtained only through annealing polycrystalline cast alloy. Materials Science and Engineering: A. 2019;**749**:249-254

[91] Tian J et al. Process optimization, microstructures and mechanical properties of a Cu-based shape memory alloy fabricated by selective laser melting. Journal of Alloys and Compounds. 2019;**785**:754-764

[92] Mazzer E, Milhorato F. Effects of aging on a spray-formed Cu-Al-Ni-Mn-Nb high temperature shape memory alloy. Materials Science and Engineering: A. 2019;**753**:232-237

[93] Adachi K, Shoji K, Hamada Y. Formation of (X) phases and origin of grain refinement effect in Cu-Al-Ni shape memory Alloys added with Titanium. ISIJ International. 1989;**29**(5):378-387

[94] Vajpai SK, Dube RK, Sangal S. Microstructure and properties of Cu-Al-Ni shape memory alloy strips prepared via hot densification rolling of argon atomized powder preforms. Materials Science and Engineering A. 2011;**529**(1):378-387

[95] Guniputi BN, Murigendrappa S. Influence of Gd on the microstructure, mechanical and shape memory properties of Cu-Al-Be polycrystalline shape memory alloy. Materials Science and Engineering: A. 2018;**737**:245-252

[96] Hussain S, Pandey A, Dasgupta R. Designed polycrystalline ultra-high ductile boron doped Cu–Al–Ni based shape memory alloy. Materials Letters. 2019;**240**:157-160

[97] Zhu M et al. Cheminform abstract: Preparation of single crystal CuAlNiBe SMA and its performances. ChemInform. 10 2009;**478**(1-2):404-410

[98] MAaSG M. Effect of heat treatment and thermal cycling on transformation temperatures of ductile Cu-Al-Ni-Mn-B alloys. Scripta Metallurgica et Materiala. 1992;**26**(11):1663-1668

[99] Roh DW, Kim JW, Cho TJ, Kim YG. Tensile properties and microstructure of microalloyed Cu-A1-Ni-X shape memory alloys. Materials Science and Engineering: A. 1991;**136**:17-23

[100] Zubair I, Khan AQ. Fascinating shape memory alloys. Journal of the Chemical Society of Pakistan. 2018:**40**(1):1-23

[101] Punburi P et al. Correlation between electron work functions of multiphase Cu-8Mn-8Al and de-alloying corrosion. Applied Surface Science. 2018;**439**:1040-1046

[102] XU, H., et al. A study on shape memory performance of single crystal Cu-Al-Ni-Be alloy. Materials Review. 2008;**4**:036

[103] Xu HP, Song GF, Mao XM. Influence of Be and Ni to Cu-Al alloy shape memory performance. Advanced Materials Research. 2011;**197**:1258-1262

[104] Gomes RM et al. Pseudoelasticity of Cu-13.8Al-Ni Alloys containing V and Nb. Advances in Science and Technology. 2008;**59**:101-107

[105] Zhu M et al. Preparation of single crystal CuAlNiBe SMA and its

performances. Journal of Alloys and Compounds. 2009;**478**(1):404-410

[106] Hodgson D, Wu MH, Biermann RJ. Metals Handbook. Vol. 2. Ohio: ASM International; 1990. p. 897

[107] Gall K et al. Thermomechanics of the shape memory effect in polymers for biomedical applications. Journal of Biomedical Materials Research Part A. 2005;**73**(3):339-348

[108] Wen C et al. Mechanical Behaviors and Biomedical Applications of Shape Memory Materials: A Review. AIMS Materials Science. 2018;**5**(4):559-590

[109] Sokolowski WM. Cold Hibernated Elastic Memory Structure: Self-Deployable Technology and Its Applications. Florida, United States: CRC Press; 2018

[110] Baghani M, Ganjiani M, Rezaei M. Numerical analysis of growing the ductile damage in structures reinforced by SMA using continuum damage mechanics approach. International Journal of Applied Mechanics. 2018;**10**(07):1850070

[111] Velmurugan C et al. Machining of NiTi-shape memory alloys-a review. Machining Science and Technology. 2018;**22**(3):355-401

[112] Leo DJ et al. Vehicular applications of smart material systems. In: 5th Annual International Symposium on Smart Structures and Materials. California, United States: International Society for Optics and Photonics; 1998

[113] Stoeckel D. Shape memory actuators for automotive applications. Materials & Design. 1990;**11**(6):302-307

[114] Rao A, Srinivasa A. A thermodynamic driving force approach for analyzing functional degradation of shape memory alloy components. Mechanics of Advanced Materials and Structures. 2018. p. 1-13. DOI: 10.1080/15376494.2018.1444229

[115] Wilkes KE, Liaw PK. The fatigue behavior of shape-memory alloys. JOM. 2000;**52**(10):45-51

[116] Sellitto A, Riccio A. Overview and future advanced engineering Applications for morphing surfaces by shape memory alloy materials. Materials. 2019;**12**(5):708

[117] Leal PB, Savi MA. Shape memory alloy-based mechanism for aeronautical application: Theory, optimization and experiment. Aerospace Science and Technology. 2018;**76**:155-163

[118] Kaya E, Kaya İ. A review on machining of NiTi shape memory alloys: The process and post process perspective. The International Journal of Advanced Manufacturing Technology. 2019;**100**(5-8):2045-2087

[119] Strittmatter J, Hiefer MEM, Gümpel P. Intelligent materials in modern production-current trends for thermal shape memory alloys. Procedia Manufacturing. 2019;**30**:347-356

[120] Hornbogen E. Review thermo-mechanical fatigue of shape memory alloys. Journal of Materials Science. 2004;**39**(2):385-399

[121] Karakoc O et al. Effects of Testing Parameters on the Fatigue Performance NiTiHf High Temperature Shape Memory Alloys. San Diego, California: AIAA Scitech, Forum; 2019. DOI. org/10.2514/6.2019-0416

[122] Gédouin P-A et al. R-phase shape memory alloy helical spring based actuators: Modeling and experiments. Sensors and Actuators A: Physical. 2019;**289**:65-76

[123] Widdle RD et al., High stiffness shape memory alloy actuated aerostructure. Google Patents; 2013

[124] Mani R, Lagoudas DC, Rediniotis OK. MEMS-based active skin for turbulent drag reduction. In: Smart Structures and Materials. California, United States: International Society for Optics and Photonics; 2003

[125] Tawfik M, Ro J-J, Mei C. Thermal post-buckling and aeroelastic behaviour of shape memory alloy reinforced plates. Smart Materials and Structures. 2002;**11**(2):297

[126] Furuya Y, Shimada H. Shape memory actuators for robotic applications. Materials & Design. 1991;**12**(1):21-28

[127] Sreekumar M et al. Critical review of current trends in shape memory alloy actuators for intelligent robots. Industrial Robot: An International Journal. 2007;**34**(4):285-294

[128] Wu MH, Schetky L. Industrial applications for shape memory alloys. In: Proceedings of the International Conference on Shape Memory and Superelastic Technologies; Pacific Grove, California. 2000

[129] Russell SM. SMST-2000: Proceedings of the International Conference on Shape Memory and Superelastic Technologies. Oxfordshire, United Kingdom: SMST, The International Organization on Shape Memory and Superelastic Technology; 2001

Chapter 4

Fabrication of Dielectric Thick Films by Electrophoretic Deposition and Their Characterization

Vineetha P and Venkata Saravanan K

Abstract

In this chapter, we have discussed about the fabrication of $MgTiO_3$ and $(K_{0.5}Na_{0.5})(Nb_{0.7}Ta_{0.3})O_3$ (KNNT) thick films via EPD, including the preparation of stable suspension, which is the essential requirement for EPD. Acetone and isopropyl alcohol are used as solvents for $MgTiO_3$ and $(K_{0.5}Na_{0.5})(Nb_{0.7}Ta_{0.3})O_3$, respectively. In both the cases triethanolamine (TEA) is used as the surfactant. It was observed that with increase in applied voltage and deposition time, the deposition of particle on the substrate has increased. For KNNT particle the optimum voltage and time to get quality film was 100 V and 10 min respectively. The surface morphology and film thickness was analyzed from SEM images. The thickness of KNNT and $MgTiO_3$ films was found to be ~32 and 18–20 μm (for film sintered at 900°C) respectively. In the case of $MgTiO_3$ thick films, the sintering temperature plays an important role in the quality of films since it affects the density and grain growth. An increase in the sintering temperature, from 800–900°C, resulted in increase in the dielectric constant and reduced loss. The dielectric values of the KNNT and $MgTiO_3$ thick film at 1 MHz was 320 and 18.3 respectively.

Keywords: electrophoretic deposition, dielectric thick film, $MgTiO_3$, $K_{0.5}Na_{0.5}Nb_{0.7}Ta_{0.3}O_3$, dielectric constant

1. Introduction

Miniaturization of electrical components have inherent advantages like less space usage, low cost, high speed and easy transportation owing to its size and light weight. In most industrial applications of piezoelectric materials, crack-free thick or thin films with a uniform microstructure are required. Lapping down a bulk material to the desired thickness is a tedious and time consuming process and thus corresponding thick films are apparently a promising alternative. Thus better understanding of the fabrication methods is necessary to make future generations of miniaturized components/materials. There are many coating techniques available to develop the materials in nano or micro sizes, which are either expensive or follow a tedious processing deposition methodology. On the other hand, preparation of thick film by the electrophoretic deposition has attained much attention owing to its high versatility, cost effectiveness, simple equipment design and process requirements [1]. Generally, thick films of any solid can be deposited by EPD provided the solid assumes its form as powder or a colloidal suspension [2]. In the following sections we have included in details the preparation of stable suspensions of KNNT and $MgTiO_3$. EPD comprises of two steps; the first step includes the preparation of a stable suspension of the material to be deposited and movement of the particle towards the oppositely charged electrode and the second step involves the deposition of the particle on the substrate surface to forms a film. After the first two deposition steps, a heat-treatment step is normally needed to further densify the deposited films and to eliminate porosity in the film [3, 4]. Compared to other advanced shaping technique, EPD offers easy control of the thickness and morphology of a deposited film through simple adjustment of the deposition time and applied potential [5].

Considering the environmental and health risks of lead-based materials, lead-based materials in the electronics industry are now being replaced with lead-free material. In lead-free ceramic materials, potassium sodium niobate-based materials have received much attention due to its superior piezoelectric properties and high Curie temperature [6]. A piezoelectric vale of 80pC/N has been reported for pure KNN ceramics prepared by solid state reaction route method [7]. However compared to the pure KNN material, doped materials have got high piezoelectric as well as ferroelectric properties. So far, studies have been carried out to improve the piezoelectric properties of KNN ceramics by partial substitutions of Li at Asite and/or Ta at B-site, [8, 9]. Tantalum doping in the B site has a major effect on the materials properties and a piezoelectric value of 205pC/N has been reported for $(K_{0.5}Na_{0.5})(Nb_{0.7}Ta_{0.3})O_3$ (KNNT) ceramics [10]. Even though the bulk ceramics such as KNNT and $MgTiO_3$ based materials have good electronic properties [10], miniaturizations of electronic devices demand smaller and lighter components with properties comparable to that of bulk material. Potassium Sodium Niobate (KNN) based layers are being employed in applications such as high frequency transducer and in ultra sound imaging [11, 12] and magnesium titanate ($MgTiO_3$) based materials are used in many applications such as high density capacitors, resonators and filters, in wireless communication systems, global positioning system, antennas for communications, radar and broadcasting satellite, operating at micro-wave and millimeter-wave frequencies [13]. The versatile use $MgTiO_3$ is due to its high quality factor (Q_f = 160,000 GHz, $Q_f = 1/\tan\delta$), low dielectric constant ($\varepsilon_r \sim 17$) and negative temperature coefficient of resonant frequency (τ_f = – 50 ppm/°C) [14].

2. Experimental

2.1 Stable suspension

In order to prepare a stable suspension of the any ceramic, it is desirable to have nano grain sized powders, hence the precursor for $MgTiO_3$, which contentions excess of MgO (Mg_2TiO_4), prepared by Solar Physical Vapor Deposition (SPVD) was used. The details of the SPVD process could be found elsewhere [15]. The main advantage of SPVD is direct synthesis of nanopowders without any contamination. Moreover, the process is eco-friendly since it uses solar energy which is available abundantly in nature. The KNNT powder required for the EPD was prepared by conventional solid state reaction technique [16, 17]. The KNNT and Mg_2TiO_4 powders were thoroughly mixed in isopropyl alcohol and acetone, respectively, followed by ultra-sonication and rigorous magnetic stirring. Ultra sonication and magnetic stirring was performed to get homogeneous and well dispersed suspensions.

In both the cases, Triethanolamine (TEA), which is an organic base, was added as the dispersant to increase the zeta potential and colloidal stability of particles in the suspension. It is known that H^+TEA species generated by proton capturing from the isopropyl alcohol/acetone are chemically adsorbed on the ceramic nanoparticles via

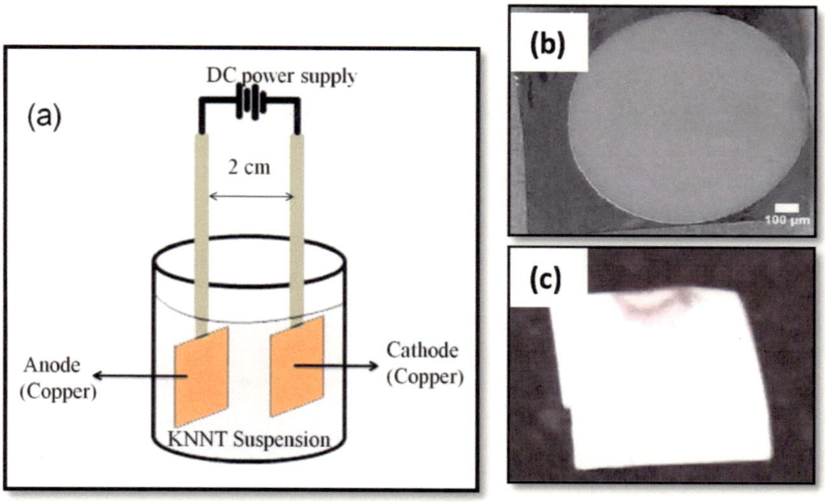

Figure 1.
(a) Schematic representation of customized EPD cell (b) optical micrograph of $MgTiO_3$ thick film [17], and (c) KNNT thick film.

strong hydrogen bonding thereby enhancing their zeta potential and hence the colloidal stability. Also, the addition of TEA is resulted in increase of pH (7 to 9 for KNNT and 6 to 8.2 for Mg_2TiO_4). The charging mechanism of KNNT in TEA added isopropyl alcohol will be discussed in the coming section. Care is taken while adding TEA since the excess amounts of TEA may lead to the sedimentation of the particle [18].

2.2 Electrophoretic deposition

The EPD was performed at room temperature using a customized EPD setup/cell (**Figure 1a**) and a voltage source (Keithely 2636 B). Pre-cleaned platinum (Pt) plates and foils are used as the electrodes. One of the peculiarities of EPD is that electrodes with various shapes and dimension can be used for deposition. The distance between the two working electrodes was kept constant at 2 cm. The deposition of KNNT was carried out at 100 V for 10 min and Mg_2TiO_4 at 60 V for 1 min. The picture of thick films obtained is shown in **Figure 1**. The deposition of KNNT as well as $MgTiO_3$ particles happened in the cathode, indicating the negative charge of particles in the suspension. The deposition of particle at cathode is known as cathodic deposition. After deposition the KNNT films were annealed in an oven at 200°C for 1 h and Mg_2TiO_4 films were dried in desiccators to evaporate the solvent. Then the KNNT films are sintered at 1000°C for 2 hr. For better compaction the dried Mg_2TiO_4 films were isostatically pressed at 150 MPa for 5 minutes before carrying out the sintering process in air at 800, 900 and 1000°C for 15 minutes in a box furnace.

X-Ray diffraction technique was used for the phase analysis of the films and the microstructures were characterized using a scanning electron microscope (SEM). Later dielectric measurements were carried-out on all the thick films for which metallic top electrodes were sputter deposited through shadow mask.

3. XRD analysis

XRD patterns of KNNT powder, green and sintered KNNT thick films are shown in **Figure 2a** and the XRD patterns of the SPVD obtained $MgTiO_3$

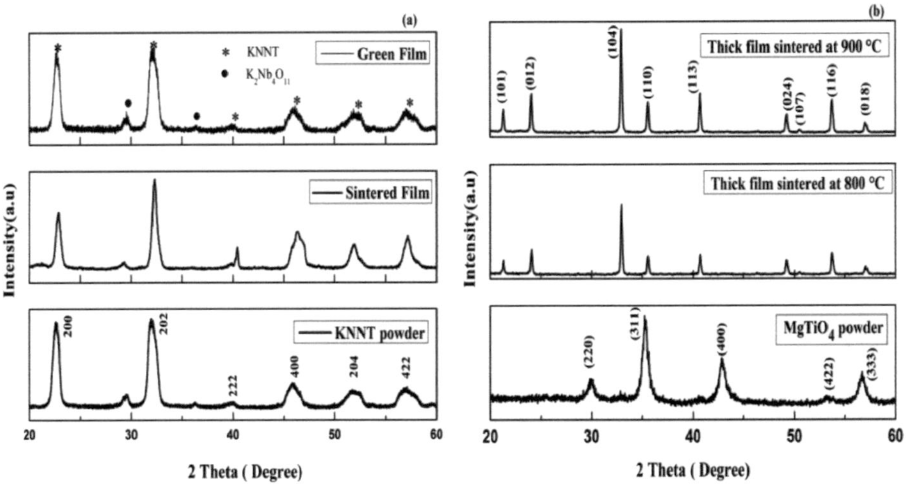

Figure 2.
(a) XRD patterns of KNNT powder, green film and sintered thick films (b) XRD patterns of Mg_2TiO_4 nanopowder, EPD films sintered at 800 (15 min) and 900°C (15 min) [17].

nanoparticle and the thick films sintered at different temperatures is shown in **Figure 2b**. XRD analysis of KNNT revealed that the powder as well as the films has perovskite monoclinic structure (JCPDS 770038) along with some secondary phase. On comparing the patterns of green and sintered films, no additional peaks were observed, indicating that the quality of the film merely depend on the quality of the powder. However, in the case of Mg_2TiO_4, a phase change between the SPVD nanopowders and the sintered films was observed. An apparent decrease in the peak width between the nanopowders and the sintered films as also observed in Mg_2TiO_4samples, which indicate an increase in the crystallite size of the sintered films [19]. Even though the starting powder was Mg_2TiO_4, a pure $MgTiO_3$ phase without secondary phases was obtained after sintering because nanopowders were very fine and very reactive. Moreover, the XRD patterns were matched to crystallographic data on synthetic geikielite crystalline phase and the peaks were indexed to a rhombohedral unit cell with a = b = 0.5057(2) nm and c = 1.3890(4) nm [17].

4. Charging mechanism and deposition

Depending on the surface charge on the particle in a suspension, the deposition can be anodic or cathodic deposition. The deposition of a negatively charged particle on anode is called anodic electrophoretic deposition and the deposition of a positively charged particle on cathode is called cathodic electrophoretic deposition. The surface charge of the particle can be easily modified by the appropriate choice of solvent or/and surfactant [18, 20]. Moreover, different surface charge can be introduced on KNN particle using a single solvent with different surfactants [21].

Here we have observed that the KNNT particles were deposited on cathode yielding a smooth deposit. Cathodic deposition of KNNT particle indicates that KNNT particles have acquired positive charge in the medium. Since TEA is a weak base, it increases the pH of the suspension and when it is added to isopropyl alcohol it takes the proton from isopropyl alcohol and acquires positive charge. The overall reaction can be written as [18].

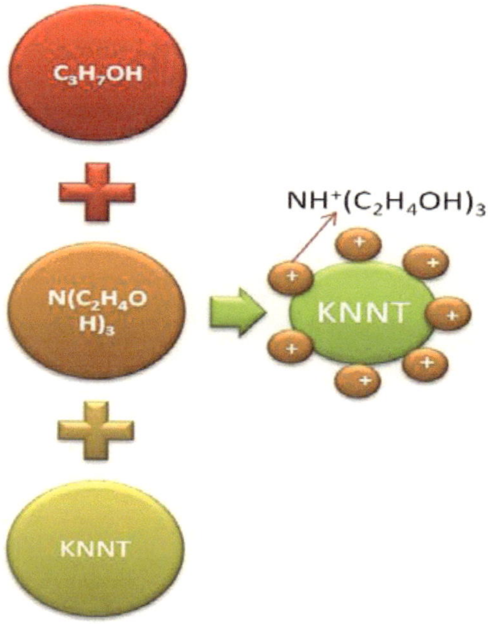

Figure 3.
Schematic representation of charge acquisition of KNNT particle in TEA added isopropyl alcohol medium.

$$C_3H_7OH + N(C_2H_4OH)_3 \leftrightarrow C_3H_7O^- + NH^+(C_2H_4OH)_3 \quad (1)$$

This charged species $NH^+(C_2H_4OH)_3$ get adsorbed on the surface of the KNNT particles confer positive charge to them and the electrostatic repulsion force between TEA adsorbed KNNT particles prevent the agglomeration of the particles in the medium. Upon applying an electric field of proper strength, the charged particles get attracted towards oppositely charged plate and there it get neutralized and the deposition of KNNT particles occurs on the plate leaving $NH^+(C_2H_4OH)_3$ back to the solution. Even though the stabilization of suspension is via electrostatic repulsion force, the high molecular weight TEA may also contribute to some steric stabilization mechanism [18]. The green and sintered film showed good adhesion with the Pt-electrode and showed no micro cracks, peelings, or pores (**Figure 3**).

5. Effect of EPD voltage and time on the deposition weight

According to Hamaker's equation the yield of electrophoretic deposition (w) is related to electrophoretic mobility (μ), surface area of the electrode (A), electric field strength (E) particle mass concentration in the suspension (C) and deposition time (t) through the following relation [22].

$$w = \int_{t_1}^{t_2} \mu . E . A . C dt \quad (2)$$

Figure 4 shows the variation of weight of KNNT particles deposited on substrate at various dc voltages and the variation in deposit weight with deposition times at a particular voltage (100 V). From **Figure 4**, it is evident that the rate of deposit increases with the increase in applied voltage and deposition time.

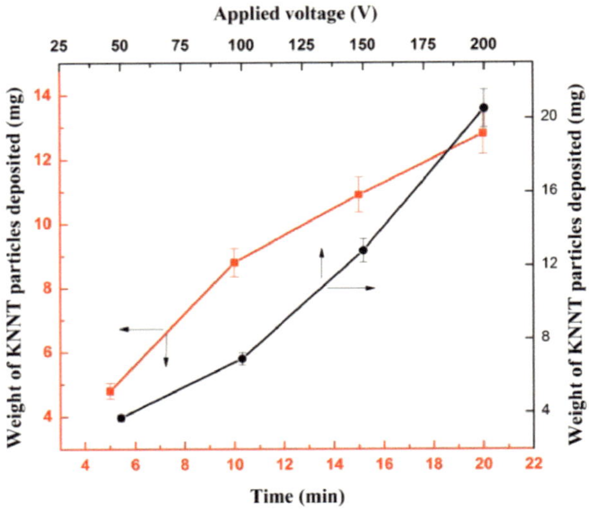

Figure 4.
Amount of KNNT particles deposited on substrate at various dc voltages and various deposition time (at constant voltage of 100 V).

At high voltage and longer deposition durations, though the rate of deposition was high, it does not yield smooth film surface. The surface morphology was very rough and the films tend to peel off from the substrate due to the low adhesion between the film and the Pt substrate. At low voltages and short deposition time, the amount of deposit was too small to form a uniformly thick film. Since we obtained films with good surface morphology at 100 V, this film was used for further studies.

6. SEM analysis

When there is dominancy of gravity and/or adhesive force of suspension over the inter particle van der Waals force, dislodgement of particle may occur during the lifting of the substrate out of suspension. This may lead to crack formation or micro cavities on the film surface. Hence, to study the quality of the deposited film, SEM images were taken and analyzed. The SEM images of KNNT thick film sintered at 1000°C are shown in **Figure 5a**. From **Figure 5a**, it is evident that KNNT particles are well deposited on the substrate with good compaction. In addition, no cracks were observed on the surface indicating that the processing conditions are favorable to get crack free thick films. Thickness as well the uniformity in thickness was analyzed from SEM images using Image J software and it was found that the film had uniform thickness of ~32 μm over the entire deposited region.

SEM micrographs $MgTiO_3$ thick film is shown in **Figure 5c–f**. A regular shaped grain with an average size of 50 nm can be observed in as deposited green films (**Figure 5c**). Whereas a porous microstructure with an average grain size of 60–70 nm was observed in $MgTiO_3$ thick films sintered at 800°C (**Figure 5d**). It is clear from **Figure 5e** that the films sintered at 900°C have more homogeneous and denser microstructure with higher grain size (100 nm). However, at higher sintering temperature (1000°C) inhomogeneous grain growth and delaminating effects were observed. The cross-sectional (CS) image of film sintered at 900°C was taken to find the thickness (inset of **Figure 5c**) and the thickness was about 18–20 μm.

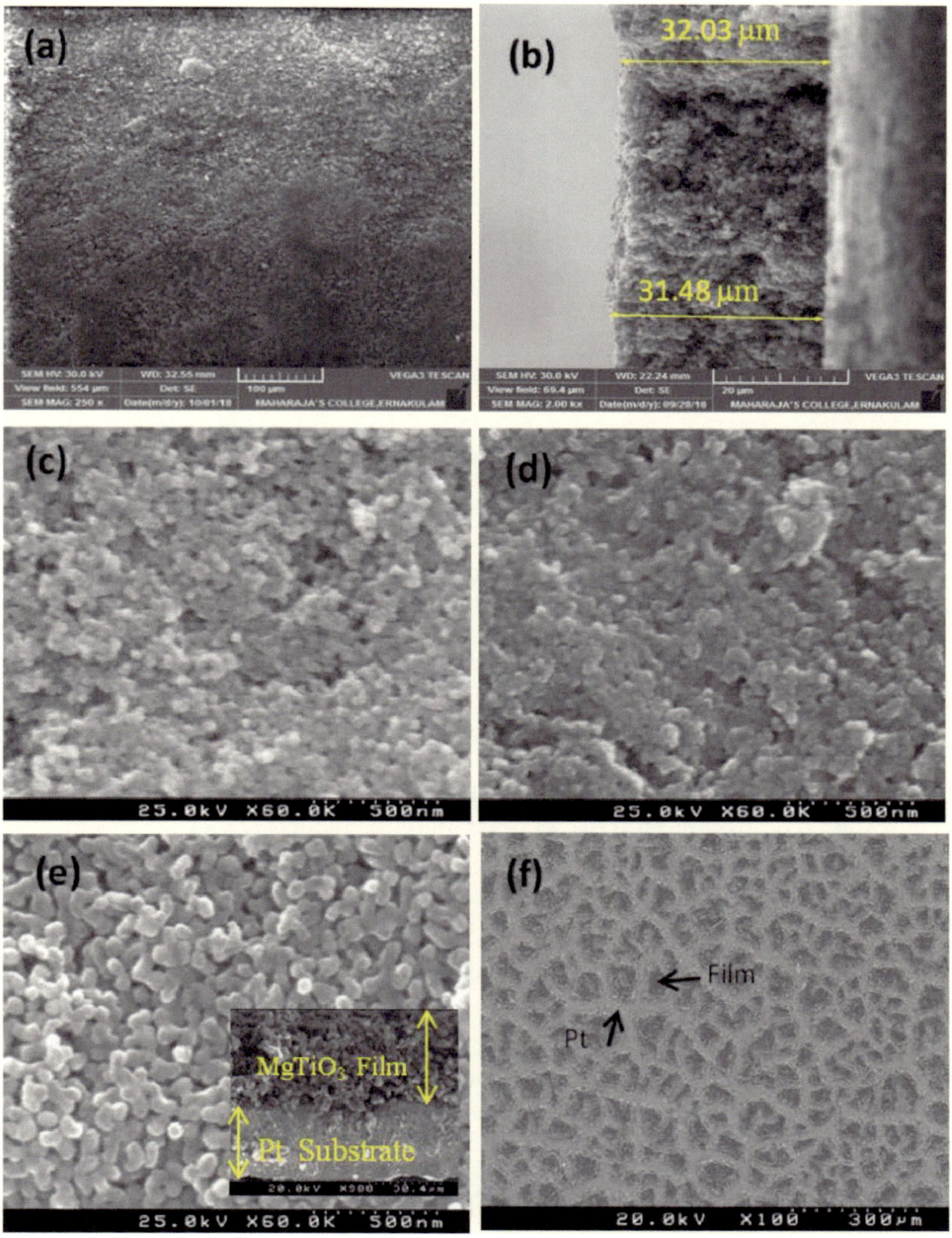

Figure 5.
SEM images (a) surface morphology of the sintered KNNT thick film (b) SEM showing the thickness of the KNNT film sintered at 1000°C. SEM micrograph of (c) as-deposited green film. (d–f) SEM images of $MgTiO_3$ thick films sintered at (b) 800°C (15 min), (c) 900°C (15 min); the inset shows the cross sectional SEM image of the film sintered at 900°C and (d) 1000°C (15 min) sintered film showing delamination (the dark island like regions are the film and the lighter area is the Pt substrate).

7. Dielectric constant (ε_r)

Variation of dielectric constant and dielectric loss with frequency at room temperature for sintered KNNT thick film is obtained using a metal- insulator-metal structure and the data is shown in **Figure 6**. Dielectric constant value of 320 (at 1 MHz) was obtained for sintered KNNT thick film. The dielectric value of the KNNT film (320) at 1 MHz was lower than the value reported for KNN thick films fabricated via electrophoretic deposition in acetone (393) [21] and aqueous

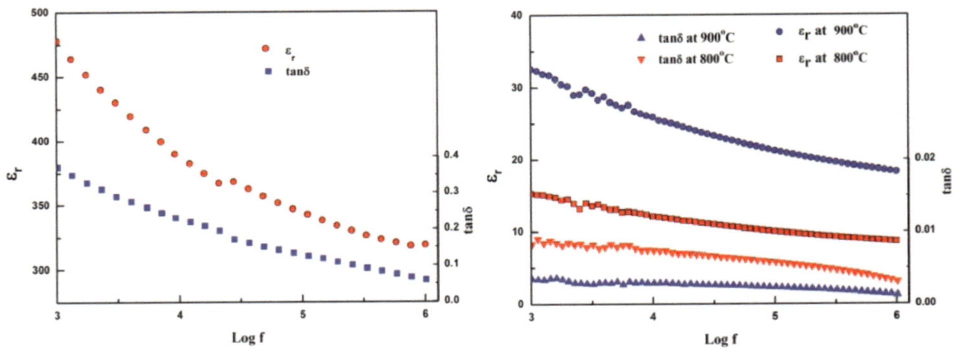

Figure 6.
Dielectric constant (ε_r) and tan δ as function of frequency for (left) KNNT thick film sintered at 1000°C (right)$MgTiO_3$ thick films sintered at 800 and 900°C.

media (495) [20] and the higher values of later two might be the result of dense film obtained by isostatic pressing and higher sintering temperature (1100°C). Dielectric permittivity and loss tangent (tan δ) of $MgTiO_3$ films were evaluated over a frequency range of 1 kHz to 1 MHz, using an impedance analyzer (HP 4284Aprecision LCR Meter). Since the samples sintered at 1000°C were not of good enough quality for electrical studies, the films sintered at 800 and 900°C were used for the studies. Among these two samples, those sintered at 900°C exhibited a good dielectric constant ~18.3 (tan δ ~ 0.0012) at 1 MHz compared to films sintered at 800°C (ε_r ~ 8.6 and tan δ ~ 0.0031). The low density and porous microstructure can be the reason for low ε_r values for the films at sintered at 800°C. ε_r value obtained for samples sintered at 900°C was comparable with that of other literature reports [23, 24].

8. Conclusions

$(K_{0.5}Na_{0.5})(Nb_{0.7}Ta_{0.3})O_3$ and $MgTiO_3$ thick films were fabricated successfully by electrophoretic deposition method from TEA added isopropyl alcohol medium and acetone medium, respectively. We suggest that the stability of the KNNT suspension is mainly due to electrostatic stabilization mechanism and the $NH^+(C_2H_4OH)_3$ ions being adsorbed on the KNNT particle is responsible for the observed positive surface charge in KNNT particles. From SEM images we confirmed that KNNT particles as well as $MgTiO_3$ particles were well deposited on the substrate and the thickness was found to be ~32 μm and 18–20 μm (for film sintered at 900°C) respectively. In the case of $MgTiO_3$ thick films, the sintering temperature plays an important role in the quality of films since it affects the density and grain growth. An increase in the sintering temperature, from 800–900°C, resulted in increase in the dielectric constant and reduced loss. Moreover, no cracks were observed on the surface of the thick films, indicating that the processing conditions are favorable to get crack free thick films. The dielectric values of the KNNT and $MgTiO_3$ thick film at 1 MHz was 320 and 18.3 respectively, which is in agreement with the earlier reported values. For these reasons we can say that compared to other solution based techniques EPD is a cost effective and simple technique to fabricate high quality thick films.

Acknowledgements

The authors acknowledge Department of Physics, Central University of Kerala for providing XRD characterization facilities and Maharajas College,

Ernakulam for SEM facility. VSK. acknowledges the facilities provided by CUTN, UGC for the start-up grant and DST-SERB for the Early Career Research grant (ECR/2015/000273). VP acknowledge the facilities and financial support offered by CUTN.

Conflict of interest

The authors declare no conflict of interest.

Author details

Vineetha P and Venkata Saravanan K*
Department of Physics, School of Basic and Applied Sciences, Central University of Tamil Nadu, Thiruvarur, India

*Address all correspondence to: venketvs@cutn.ac.in

References

[1] Corni I, Ryan MP, Boccaccini AR. Electrophoretic deposition: From traditional ceramics to nanotechnology. Journal of the European Ceramic Society. 2008;**28**:1353-1367. DOI: 10.1016/j.jeurceramsoc.2007.12.011

[2] De Riccardis MF. Ceramic Coatings-Applications in Engineering. China: InTech; 2012. 286p

[3] Sarkar P, Nicholson PS. Eletrophoretic deposition (EPD): Mechanisms, kinetics and applications to ceramics. Journal of the American Ceramic Society. 1996;**79**:1987-1202. DOI: 10.1111/j.1151-2916.1996.tb08929.x

[4] Besra L, Liu M. A review on fundamentals and applications of electrophoretic deposition (EPD). Progress in Materials Science. 2007; **52**:1-61. DOI: 10.1016/j.pmatsci.2006.07.001

[5] Heavens N. Electrophoretic deposition as a processing route for ceramics. In: Binner GP, editor. Advanced Ceramic Processing and Technology. Vol. 1. Park Ridge, NJ, USA: Noyes Publications; 1990. pp. 255-283

[6] Wu J, Xiao D, Zhu J. Potassium–sodium niobate lead-free piezoelectric materials: Past, present, and future of phase boundaries. Chemical Reviews. 2015;**115**:2559-2595. DOI: 10.1021/cr5006809

[7] Egerton L, Dillon DM. Piezoelectric and dielectric properties of ceramics in the system potassium-sodium niobate. Journal of the American Ceramic Society. 1959;**42**:438. DOI: 10.1063/1.2436923.

[8] Ming et al. Piezoelectric properties of (Li, Sb, Ta) modified (Na, K) NbO_3 lead-free ceramics. Journal of Applied Physics. 2007;**101**:054103. DOI: 10.1063/1.2206554

[9] Zang et al. Perovskite lead-free $(Na_{0.5}K_{0.5})_{1-x}(LiSb)_xNb_{1-x}O_3$ piezoceramics. Applied Physics Letters. 2006;**88**:212908

[10] Lv YG et al. Tantalum influence on physical properties of $(K_{0.5}Na_{0.5})(Nb_{1\_x}Ta_x)O_3$ ceramics. Materials Research Bulletin. 2009;**44**:284-287. DOI: 10.1016/j.materresbull.2008.06.019

[11] Lau ST, Li X, Zhou QF, Shung KK, Ryu J, Park D-S. Aerosol-deposited KNN–LSO lead-free piezoelectric thick film for high frequency transducer applications. Sensors and Actuators A: Physical. 2010;**163**:226-230. DOI: 10.1016/j.sna.2010.08.020

[12] Zhu et al. (100)-textured KNN-based thick film with enhanced piezoelectric property for intravascular ultrasound imaging. Applied Physics Letters. 2015;**106**:173504. DOI: 10.1063/1.4919387

[13] Ferri EAV, Sczancovski JC, Cavalcante LS, Paris EC, Espinosa JWM, de Figueiredo T, et al. Photoluminescence behavior in $MgTiO_3$ powders with vacancy/distorted clusters and octahedral tilting. Materials Chemistry and Physics. 2009;**117**:192-198. DOI: 10.1016/j.matchemphys.2009.05.042

[14] Istianah, Lestari R, Baqiya MA, Pratapa S. In: Proceedings of International Conference on X-Ray Microscopy and Smart Matter (ICXSM 2011), June 13, 2011; Solo, Indonesia. 2011. pp. 1-4

[15] Apostol I, Saravanan KV, Monty CJA, Vilarinho PM. Solar physical vapor deposition: A new approach for preparing magnesium titanate nanopowders. Applied Surface Science. 2013;**285**:49-55. DOI: 10.1016/j.apsusc.2013.07.155

[16] Vineetha P, venkata Saravanan K. Fabrication of $(K_{0.5}Na_{0.5})(Nb_{0.7}Ta_{0.3})O_3$ thick films by electrophoretic deposition. AIP Conference Proceedings. 2018;**1953**:090041

[17] Apostol I, Mahajan A, Monty CJA, Venkata Saravanan K. Nanostructured $MgTiO_3$ thick films obtained by electrophoretic deposition from nanopowders prepared by solar PVD. Applied Surface Science. 2015;**358**:641-646. DOI: 10.1016/j.apsusc.2015.09.060

[18] Farrokhi-rad M, Ghorbani M. Stability of titania nano-particles in different alcohols. Ceramics International. 2012;**38**:3893-3900. DOI: 10.1016/j.ceramint.2012.01.041

[19] Monty CJA. Characterization and properties of nanophases prepared by solar physical vapor deposition (SPVD) in the solar reactor heliotron. Arabian Journal for Science and Engineering. 2010;**35**(1C):93-116

[20] Mahajan A et al. Unleashing the full sustainable potential of thick films of lead-free potassium sodium niobate ($K_{0.5}Na_{0.5}NbO_3$) by aqueous electrophoretic deposition. Langmuir. 2016;**32**:5241-5249. DOI: 10.1021/acs.langmuir.6b00669

[21] Dolhen M et al. Sodium potassium niobate ($K_{0.5}Na_{0.5}NbO_3$, KNN) thick films by electrophoretic deposition. RSC Advances. 2015;**5**:4698-4706. DOI: 10.1039/c4ra11058g

[22] Hamaker HC. Formation of a deposit by electrophoresis. Transactions of the Faraday Society. 1940;**36**:279-283. DOI: 10.1039/TF9403500279

[23] Kim ES, Seo SN. Evaluation of microwave dielectric properties of $MgO-TiO_2$ system by dielectric mixing rules. Journal of the Korean Ceramic Society. 2010;**47**:163-168. DOI: 10.4191/KCERS.2010.47.2.163

[24] Wu HT, Jiang YS, Cui YJ, Zhang XH, Jia X, Yue YL. Improvements in the sintering behavior and microwave dielectric properties of geikielite-type $MgTiO_3$ ceramics. Journal of Electronic Materials. 2013;**42**:445-451. DOI: 10.1007/s11664-012-2349-2

Chapter 5

Effect of Electrolytic Compositions in Kinetics Mechanism of High-Purity Titanium Electrochemical Extraction Process

Jianxun Song, Yusi Che, Yongchun Shu and Jilin He

Abstract

High pure titanium, which is a critical material used for integrated circuit (IC) manufacturing, can be extracted by molten salt-electrolysis process. It will be widely used in the future for the process, is simple and easy, to achieve continuous production. However, some scientific questions need to be clarified at present. (1) Current efficiency needs to be enhanced by way of electrolyte designing and selection; (2) Product quality needs to be improved by means of electrolyte purification; (3) Electrolytic parameters need to be optimized for obtaining a better morphology. Above on, this chapter aims to explore the behaviors of titanium ions in various molten salts by means of chemical analysis and electrochemical testing. The complexes will be discussed for clarifying the influence of electrolytic compositions on kinetics mechanisms of the electrolysis process and the properties of the molten salt. A quantitative method for estimating oxygen content will be also discussed for the purpose of optimizing the composition of electrolytes. The chapter will provide a better understanding mechanism of kinetics of high pure titanium electrolysis, and the basic theory and experimental data can be used for reference in the industrialization process.

Keywords: electrolysis, molten salt structures, kinetics mechanism, high-purity titanium

1. Introduction

High-purity titanium has excellent properties, such as light weight, corrosion resistance, and low electrical resistivity [1, 2]. It is mainly used in large-scale integrated circuit (IC) manufacturing, new type of titanium-alloy and other fields. Therefore, it is a strategic material for electronic and aerospace.

As a sputtering target for manufacturing large-scale IC, a purity of 4 N5 (99.995%) to 6 N (99.9999%) is required for titanium, in which the oxygen content is below 200 ppm, and the metal impurity content is less than 10 ppm. However, the current technology for preparing high-purity titanium has caused a high price, which directly leads to lacking of application in large scale. Thereby, the development of new methods and technologies to prepare high-quality and low-cost

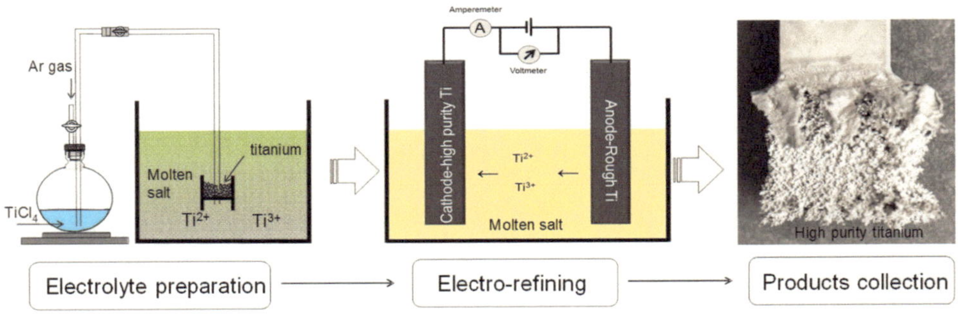

Figure 1.
Diagram of electro-refining process of high-purity titanium in molten salt.

titanium will greatly promote large-scale IC industry. Considering the preparation technology of high-purity titanium, the process of electro-refining in molten salt which is simple in process and easy to realize successive has a very broad application prospect, and has attracted much attention. The extraction process of high-purity titanium in molten salt is shown in **Figure 1**.

In the process, for the first step, $TiCl_4$ and high-purity titanium were used to react for obtaining a molten salt containing low-valent titanium ions (Ti^{2+} and Ti^{3+}). The salt will be used as a source of titanium ions to be added to the eutectic salt. Second, titanium sponge which comes from Kroll process is used as anode, and a high-purity titanium plate is used for cathode. During the working of electrolysis, the anode is electrochemically dissolved and the electrochemical deposition of high-purity titanium occurs on the cathode. Finally, high-purity titanium products with different morphology and quality are obtained under different electrolysis parameters.

At present, the problems that need to be solved in the industrialization process of high-purity titanium extraction with the method of electro-refining in molten salt are below: (a) current efficiency need to be improved; (b) product qualities should be enhanced. Since the existence of various state of titanium ions in the molten salt (Ti^{2+}, Ti^{3+}, and Ti^{4+}), the disproportionation reactions between titanium ions and the metallic titanium cause a current consumption. Thus, it is critical to reveal the behavior of titanium ions in different electrolyte compositions/structures for improving current efficiency. In addition, oxygen has a great influence on the quality of the product, and the oxygen ions in the electrolyte cannot be completely removed. It is of great practical significance to discuss the influence of oxygen ions on the structure of the molten salt for enhancing the product quality. With a better understanding of the kinetics in titanium electrolysis process on the basis of systematic investigation of electrolyte microstructures, it will provide theoretical guidance for high-purity titanium electro-refining process in molten salt, and has important reference value for its industrialization test.

In this chapter, it aims to explore the behaviors of titanium ions in various molten salts by means of chemical analysis and electrochemical testing. The complexes will be discussed for clarifying the influence of electrolytic compositions on kinetics mechanisms of the electrolysis process and the properties of the molten salt.

2. Experimental methods

2.1 Electrochemical analysis

The electrolyte was melted in the alumina crucible after being placed in a furnace. The electrochemical analyses, such as cyclic voltammetry (CV), square wave voltammetry (SWV), line scanning voltammetry (LSV), and so on, were performed in an inert argon atmosphere previously dehydrated by heating in a vacuum. All measurements were carried out under the atmosphere of dried argon. In order to investigate the influence of the radius of the cations and fluoride ion in the melts, analyses were executed in various types of molten salts.

All the electrochemical tests were carried out in a three-electrodes configuration with a glassy carbon or metallic rod/desk being the working electrode, a spectral graphite rod being counter electrode, and an Ag/AgCl electrode or metallic titanium rod being the reference electrode, respectively. Sometimes, the potential of the reference electrode was calibrated relative to chlorine gas emission by using galvanic square wave technology.

In addition, during the electrochemical testing, the gas was analyzed on line with a resolution mass spectra and a high-sensitivity gas analysis instrument connected to the electrode tail tubing. The gas-collecting device was made with a quartz bell jar, and it correlates with resolution mass spectra, and high-purity helium was used as a carrier gas.

2.2 Concentration determination

A special quartz sampler consisting of an injector and a quartz tube was applied for determining the concentrations of Ti^{2+} and Ti^{3+} ions. The injector on top of the quartz tube is sealed with a rubber plug. Four parallel samples were extracted from the molten salts for analysis. The concentrations of Ti^{2+} and Ti^{3+} ions in the sample were determined by H_2 volumetric analysis and titration, respectively. According to reaction (1), the concentration of Ti^{2+} was quantified by H_2 volumetric analysis.

$$2Ti^{2+} + 2H^{+} = 2Ti^{3+} + H_2\ (g) \tag{1}$$

It is worth noting that that the dissolved oxygen in an aqueous solution will oxidize titanium ions from Ti^{2+} into Ti^{3+}, which is also the reason why the concentration of Ti^{2+} ions is underestimated. Thus, the deionized water was treated by vacuum degassing and completely cleaned by the high-purity argon gas. A specific amount of concentrated hydrochloric acid was injected into the deoxygenized water to prepare diluted hydrochloric acid (1 mol/L). The deoxygenized hydrochloric acid solution was saturated by bubbling high-purity H_2 for 30 min to prevent the evolution of H_2 by reaction (1).

Note that the concentration of Ti^{3+} ions in the solution consisted of the initial Ti^{3+} ions in the sample and the oxidized Ti^{3+} ions. The total concentration of Ti^{3+} ions in the solution was determined by titration using 0.05 mol/L $NH_4Fe(SO_4)_2$ aqueous solution. The Ti^{3+} in the solution reacted with Fe^{3+} follow by reaction (2).

$$Ti^{3+} + Fe^{3+} = Ti^{4+} + Fe^{2+} \tag{2}$$

Finally, the concentration of Ti^{4+} from reaction (2) was determined by diantipyryl methane spectrophotometry, the concentration of Ti^{4+} was equal to the concentration of Ti^{2+} plus Ti^{3+}.

2.3 Electrochemical titration technique

The concentration of O^{2-} in the molten salt was determined by the $P_O{}^{2-}$ indicator electrode consisted of a tube of Yttria Stabilized Zirconia (YSZ), and the reference electrode was an Ag/AgCl electrode. The auxiliary electrode was a spectrum pure graphite rod. The variation of the YSZ electrode potential was recorded using the potentiostat.

The Nernstian behavior of the system was demonstrated by measuring the change in electrode potential. The relationship between the concentration of oxide and potential can be expressed as Eq. (3):

$$E_{O^2/O^{2-}} = E^*_{O^2/O^{2-}} - \frac{RT}{2F} P_{O^{2-}} \quad (3)$$

where $E_{O2/O^{2-}}$ is the equilibrium potential of O_2/O^{2-} system (in V); $E^*_{O2/O^{2-}}$ is an apparent standard potential of the system (in V) and $P_{O^{2-}}$ is ($-\ln x_O{}^{2-}$); and $x_{O^{2-}}$is the concentration of O^{2-}.

3. Results and discussion

3.1 Influence of electrolyte cation on ionic equilibrium of titanium

Metallic titanium and titanium ions undergo the following disproportionation reactions in the molten salt, and their equilibrium constants can be expressed by Eqs. (6) and (7), respectively [3, 4].

$$3Ti^{2+} = Ti + 2Ti^{3+} \quad (4)$$

$$4Ti^{3+} = Ti + 3Ti^{4+} \quad (5)$$

$$K_{c1} = \frac{x^2{}_{Ti^{3+}} \cdot x_{Ti}}{x^3{}_{Ti^{2+}}} = \frac{x^2{}_{Ti^{3+}}}{x^3{}_{Ti^{2+}}} \quad (6)$$

$$K_{c2} = \frac{x^3{}_{Ti^{4+}} \cdot x_{Ti}}{x^4{}_{Ti^{3+}}} = \frac{x^3{}_{Ti^{4+}}}{x^4{}_{Ti^{3+}}} \quad (7)$$

Titanium ions with various valent states will compete with host cations, such as alkali metal or alkaline earth metal ions when they are used as a cation to coordinate with a host anion (such as Cl^-) in molten salt. The coordination rules between titanium ions and anions will affect the direction of the above equilibrium reactions, thereby the existence of titanium ions in the molten salt will be changed. Moreover, the electrolysis process will be a one-step reduction of Ti^{2+} to Ti when titanium ions in the molten salt are stably present in form of Ti^{2+}. On the other hand, when Ti^{3+} or higher-valent titanium ions are stabilized, the electrolysis process will achieve a one-step reduction of Ti^{3+} to Ti. **Table 1** shows the reduction potentials of titanium ions in different chloride molten salts.

Results show that the reduction of titanium ions in above molten salts has undergone a two-step process of $Ti^{3+} \rightarrow Ti^{2+} \rightarrow Ti$ except CsCl-LiCl [5]. It can also be concluded that the smaller radius of the alkali metal or alkaline earth metal cation in the electrolyte, the stronger the coordination ability with the anion (Cl^-). Considering the mechanism in deep-layer, it can be interpreted that it is impossible to have enough Cl^- around Ti^{3+} in a molten with a smaller radius of the electrolyte. Then, the instability of Ti^{3+} in reaction (4) causes the reaction to move to the left, and the concentration of Ti^{2+} increases in the total concentration of titanium ions. When the average valence state of titanium ions in the molten salt is around 2.2–2.5, the electrolytic reduction of Ti^{3+} needs to undergo a process of $Ti^{3+} \rightarrow Ti^{2+} \rightarrow Ti$. On the contrary, when the average radius of electrolyte cation is larger, such as CsCl, Ti^{3+} is relatively stable, and the average valence state of titanium ions is about 2.7–2.9, which can realize one-step reduction of Ti^{3+} to Ti.

Molten salt	$E_{Ti^{3+}/Ti^{2+}}$, V	$E_{Ti^{2+}/Ti}$, V	$E_{Ti^{3+}/Ti}$, V	Ref. electrode/Temp.
$CaCl_2$	0.20 [6]	−0.10 [6]	—	Ti, 1123 K [6]
$CaCl_2$-NaCl (1:1)	−0.85 [7]	−1.20 [7]	—	Ag^+/Ag, 823 K [7]
LiCl-KCl (59:41)	−1.05 [8]	−1.68 [8]	—	Ag^+/Ag, 723 K [8]
LiCl-KCl (59:41)	−1.78 [9]	−1.9 0 [9]	—	Cl_2/Cl^-, 993 K [9]
CsCl-LiCl (59.5:40.5)	−1.88 [9]	−1.93 [9]	—	Cl_2/Cl^-, 993 K [9]
CsCl-LiCl (59.5:40.5)	—	—	−2.01 [9]	Cl_2/Cl^-, 793 K [9]
NaCl-KCl (1:1)	−1.75 [10]	−2.05 [10]	—	Cl_2/Cl^-, 1023 K [10]
	0.18 [11]	−0.18 [11]	—	Ti, 1023 K [11]
	−1.95 [9]	−1.98 [9]	—	Cl_2/Cl^-, 993 K [9]
	0.06 [12]	−0.06 [12]	—	Ti, 1006 K [12]
NaCl-CsCl (1:2)	−1.55 [13]	−1.80 [13]	—	Cl_2/Cl^-, 873 K [13]

Table 1.
Reduction potentials in various chloride molten salt.

In order to deliberate the effect of cation in electrolyte, the relationship between K_c and cation radius was studied. It should be noted that the K_c are the final results and were analyzed by best-fitting method [14].

It was studied that the stability of chlorocomplexes is strongly affected by different solvent cations. One explanation may be described by a simple electrostatic model. Thus, it can be deduced that the titanium ions at their higher oxidation states are engaged in chlorocomplexes such as $TiCl_4^-$ or $TiCl_6^{3-}$ from the model. Moreover, there is a competition between alkali metal and titanium ions to associate with chloride ions. The smaller ionic radius of the alkali metal ions is, the more strongly associate with Cl^-, which results in a lesser stability of chlorocomplexes, and the stability of the Ti^{2+} complex increased in the presence of outer-sphere cations in the series Cs^+ to Li^+. Under the circumstances, the maximum Ti^{2+} complex stability is achieved in the presence of Li^+ outer sphere cations.

For mixtures of alkali chlorides, the polarizing ability of alkali cations to Cl^- could be an average of each according to the mix ratio. Polarizing power is introduced for describing the numerous semi-quantitative relationships, involving the charge and the radius of the solvent cations can be present as follows:

$$P = \sum x_i \frac{Z_i}{r_i^2} \tag{8}$$

where the suffix *i* represents the solvent cation, Z_i is the charge of the cation, and r_i is the radius of it. The relationships between equilibrium constant and polarizing power is shown in **Figure 2**.

The values of K_c decreased with the enlargement of the polarizing power. The results also showed that the Ti^{3+} is less stable in the alkali chloride with a stronger polarizing power. A larger polarizing power of electrolyte results in a lesser stability of $TiCl_6^{3-}$ and smaller K_c value. The stronger polarizing power is the closer of Cl^- combined with the alkali cation. More importantly, the K_c is a constant value when polarizing power is constant regardless in KCl-LiCl or CsCl-LiCl at 1023 K.

From the above, it can be concluded that the cation of electrolyte will impact on the equilibrium of disproportionation reactions (4) and (5). The reduction steps were revealed in various molten salt through electrochemical method, and the competition relationship between cation and titanium ion was also proved. Hence,

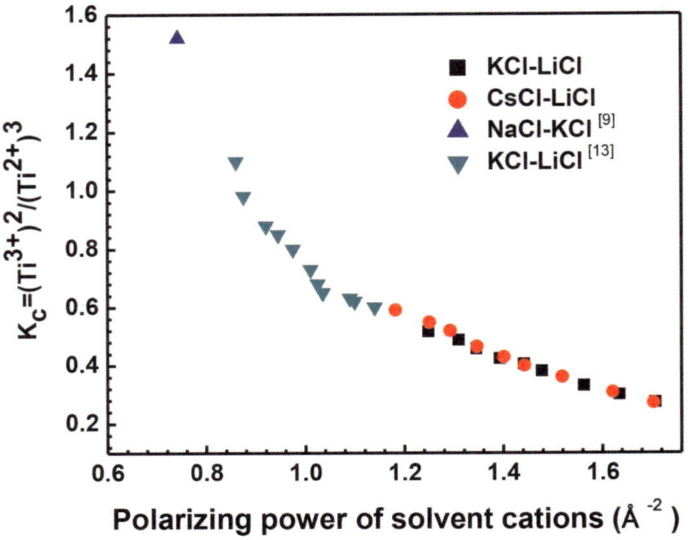

Figure 2.
The relationship between polarizing power of electrolyte and equilibrium constant at 1023 K (reprinted from Ref. [14]).

selecting one type of chloride molten salt in the high-purity titanium electrorefining process has great directive significance.

3.2 Influence of electrolyte anion on ionic equilibrium of titanium

3.2.1 Effect of fluoride ion on the state of titanium ions in molten salt

Relative to chloride ions, the ionic radius of fluoride ions is smaller, and it is easier to form a coordination relationship with the cations in the molten salt [14–19]. Therefore, coordination compounds, TiF_i^{n+}, will be formed between fluorine ions and titanium ions in a fluoride containing molten salt. The formation of these coordination compounds will effect on the valence equilibrium of titanium ions in the molten salt. **Figure 3** shows the variation of the apparent equilibrium constant of disproportionation

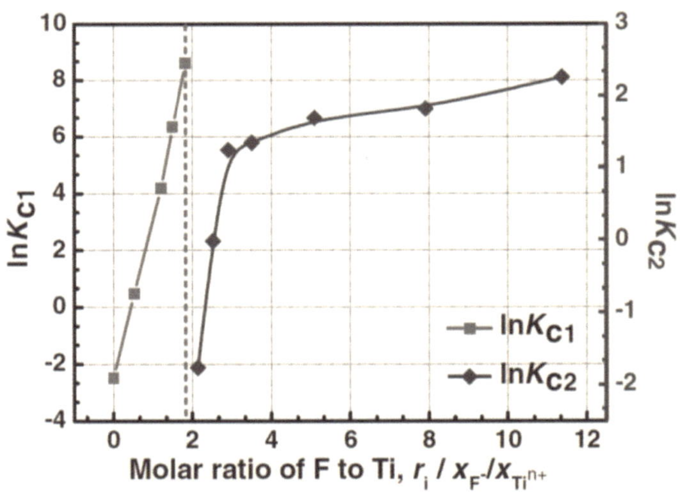

Figure 3.
Relationship between r_i *and apparent equilibrium constant of reaction (4) and (5) in NaCl-KCl molten salt at 1023 K (reprinted from Ref. [18]).*

reactions (4) and (5) in NaCl-KCl molten salt under various content of fluoride [18]. It has been shown that the apparent equilibrium constants (K_{c1} and K_{c2}) for reactions (4) and (5) increased with the increasing of fluoride ion concentration in the molten salt. The equilibrium relationship between titanium ions and titanium metal is mainly present in the reaction (4) when the ratio of fluorine to titanium (r_i) is less than 1.8. $Ti^{2+}$was not detected in the molten salt with r_i greater than 1.8. Thus, the titanium ions in the molten salt were transferred, and it is present in reaction (5).

r_i is an important parameter to qualitatively investigate the equilibrium reactions among titanium ions. The initial concentration of titanium ions is also a factor that cannot be ignored. **Figure 4** is a graph showing the color change of samples under different r_i conditions at an initial concentration of Ti^{3+} of 4.6×10^{-3} (mol %). It can be seen that the color of the molten salt gradually changed from emerald green (color of Ti^{2+}) to the deep red (color of Ti^{3+}), and then gradually become shallower at various concentrations of fluoride ion.

The relationship between the initial concentration of Ti^{3+}, F^- and the concentration of the coordination compounds in the molten salt is shown in **Figure 5**. It can be seen that these three parameters are a smooth surface on the three-dimensional map. It indicates that the total concentration of the coordination compound is determined by the two parameters: initial concentration of Ti^{3+} and F^-.

The coordination reaction between fluoride ion and titanium ion (Ti^{3+}) is as shown in reaction (9):

$$Ti^{3+} + i\,F^- = TiF_i^{3-i}, i = 1 \sim 6 \tag{9}$$

The equilibrium constants of reaction (9) were evaluated using Eq. (10) [19]. **Table 2** shows the results of the best fitting of equilibrium constants for the complex formation.

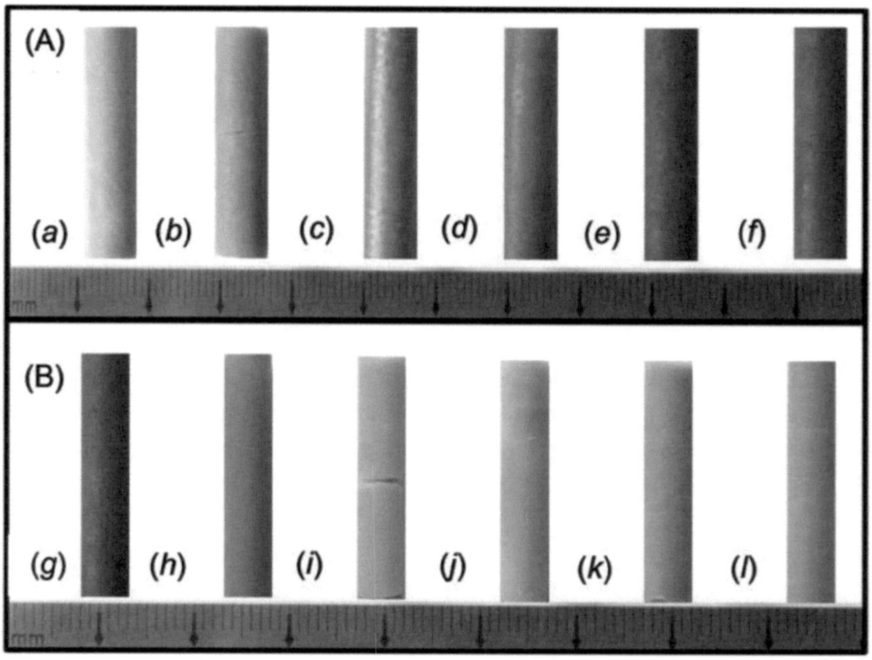

Figure 4.
Samples of titanium chloride in the NaCl-KCl melt with different concentrations of fluoride at 1023 K. the initial concentration of Ti(III) was 0.46×10^{-2} and the concentration of fluoride, $x_{totalF^-}/10^{-2}$, was (A) region of a low fluoride content: (a) 0.0; (b) 2.6; (c) 5.3; (d) 6.4; (e) 7.7; (f) 9.0; (B) region of a higher fluoride content: (g) 10.0; (h) 10.5; (i)11.4; (j) 12.1; (k) 13.0; and (l) 13.8 (reprinted from Ref. [19]).

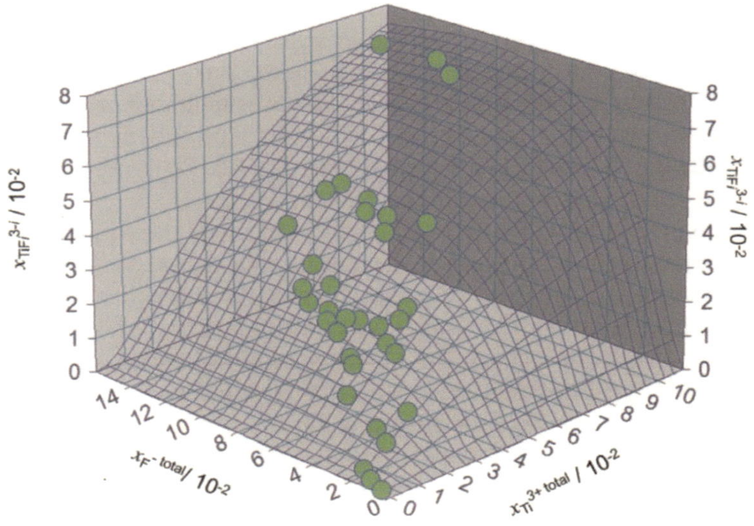

Figure 5.
The relationship between concentration of fluoride, Ti(III) and concentration of complexes at 1023 K (reprinted from Ref. [19]).

TiF_i^{3-i}	K_i	ΔG_i^0/kJ mol^{-1}	$\Delta G_i^0 i^{-1}$/kJ mol^{-1}
$i = 1$	3.9	−11.6	−11.6
$i = 2$	3.5×10^2	−49.8	−24.9
$i = 3$	4.5×10^{-5}	85.2	28.4
$i = 4$	4.1×10^4	−90.4	−22.6
$i = 5$	1.6×10^{-8}	133.1	26.6
$i = 6$	5.8×10^6	−132.4	−22.1

Table 2.
The formation equilibrium constant and Gibbs free energies of complexes formation (reprinted from Ref. [19]).

$$K_i = \frac{a_{TiF_i^{3-i}}}{\left(a_{F^-}^{eq.}\right)^i a_{Ti^{3+}}^{eq.}} = \frac{x_{TiF_i^{3-i}}}{\left(x_{F^-}^{eq.}\right)^i x_{Ti^{3+}}^{eq.}}, i = 1 \sim 6 \quad (10)$$

Thus, it was found that the formation equilibrium constants of the complexes, TiF_i^{3-i} (i = 1~6), can be revalued by the best-fitting method. They are 3.9, 3.5×10^2, 4.5×10^{-5}, 4.1×10^4, 1.6×10^{-8}, and 5.8×10^6, respectively. Using the equilibrium constants, the Gibbs free energies for the complex formation were revealed by Eq. (11):

$$\Delta G_i{}^0 = -RT\ln(K_i)\ (i = 1 \sim 6) \quad (11)$$

where R is the gas constant and T is the absolute temperature. The results of the Gibbs free energies for the complex formation reactions are also shown in **Table 2**. It can be seen that the values of $\Delta G_1{}^0$, $\Delta G_2{}^0$, $\Delta G_4{}^0$, and $\Delta G_6{}^0$ are negative, while $\Delta G_3{}^0$ and $\Delta G_5{}^0$ is positive. This corresponds to the extremely small value of K_3 and K_5. It is well understandable in the coordination theory that the way of asymmetric coordination is uncommon. The real coordination number of anion (Cl^-, F^-) around each Ti(III) is 6, i is the number of F^- ion replaced the position of Cl^-.

In the complex with 3 and 5 F^-, one of F^- is coordinated in the opposite position of Cl^-. This asymmetric structure will cause relatively higher chemical potential than a symmetric structure. Therefore, the mainly existential states of the complexes are TiF^{2+}, TiF_2^+, TiF_4^-, and TiF_6^{3-} in NaCl-KCl melt when fluoride ions were added. The value of $\Delta G_i^0/i$ is also shown in **Table 2**. This value corresponds to the change of the Gibbs free energies for the reactions of each F^-replacing the position of Cl^-. As can be found from the table, the values are very close to each other, for TiF_2^+, TiF_4^-, and TiF_6^{3-}, which is reasonable because they are all in symmetric coordination.

3.2.2 Effect of fluoride ion on reduction steps of titanium ions

The chemical behavior of fluoride ion and titanium ion in electrochemical process was examined by chemical method, respectively. The influencing mechanism of fluoride ion on the redox process of titanium ion was determined by various electrochemical analysis methods [20]. Under the condition of fluoride free in molten salt, cyclic voltammetry scanning results show that the Ti^{3+} reduction process was proceeded two-step, and they are below: $Ti^{3+} \rightarrow Ti^{2+}$ and $Ti^{2+} \rightarrow Ti$, respectively. However, when r_i was equals to 3.0, the reduction of Ti^{3+} was achieved by one step of $Ti^{3+} \rightarrow Ti$. Under such conditions, the power loss caused by the disproportionation reaction (4) can be effectively reduced.

The above studies show that fluoride ions can form coordination compounds with titanium ions to change the micro-structure of the molten salt, affect the equilibrium movement and electrochemical reduction steps of titanium ions in molten salt.

3.3 Influence of oxygen ions on ionic equilibrium of titanium

3.3.1 The formation of TiOCl in molten salt

Normally, the equilibrium constants of the disproportionation reactions were determined by concentration analyzing. However, the obtained equilibrium constant value (K_{c1}) for reaction (4) is severely dissipated, as shown in **Figure 6**.

The volatility of the K_{c1} value in molten salt that evaluated by the HCl treated is relatively small. Sekimoto [21] and other studies have shown that this is due to the coordination reaction (12) between Ti^{3+} and O^{2-} in the molten salt. The removal of oxygen from the molten salt by HCl will eliminate the effect of TiOCl on the concentration determination to some extent.

$$Ti^{3+} + O^{2-} + Cl^- = TiOCl \quad (12)$$

Based on above research, the influence of oxygen ions on the evaluation of the equilibrium constant of disproportionation reaction was considered, and the equilibrium constant value of reaction (4), the solubility product (K_{sp}) of TiOCl were estimated by the best-fitting method. More importantly, it provides a reliable method for evaluating the equilibrium constant of the disproportionation reaction. By using this method, a systematically study was carried out to investigate the influence of cation of electrolyte on the equilibrium constant [14, 23].

Considering the electrochemical behavior of Ti-O ions, Barner speculated that the intervention of oxygen ions makes the electrochemical reduction process similar to the homogeneous reaction due to the presence of the coordination compounds TiF_6^{2-} and $TiOF_5^{3-}$ [24].

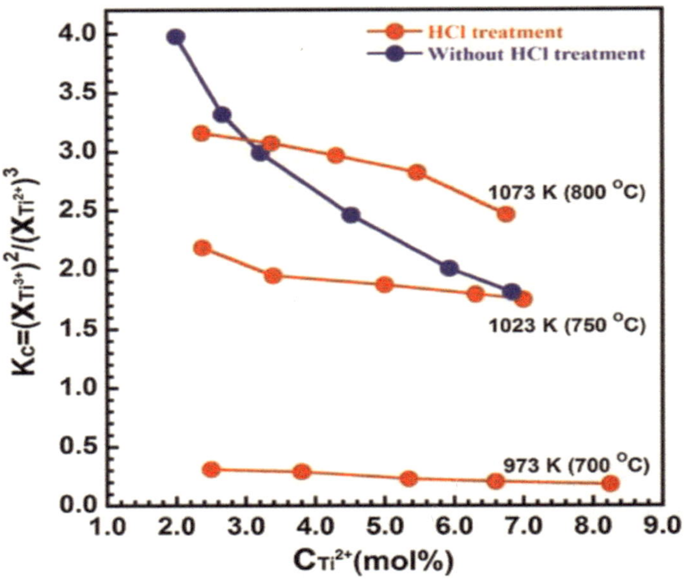

Figure 6.
Experiment data of K_{c1} and $C_{Ti}{}^{2+}$ in NaCl-KCl equimolar melt at 973, 1023, and 1073 K, respectively (reprinted from Ref. [22]).

The above studies have shown that the formation of a coordination compound between oxygen ions and high-valent titanium ions in the molten salt brings about a deviation in the evaluation of the concentration of titanium ions. At the same time, Ti-O coordination compounds also affect the electrochemical reduction process of titanium ions.

3.3.2 Purification and quantification of oxide content

Oxygen ions are very sensitive elements in the high-purity titanium electrolytic refining process, so it is necessary to find a simple and accurate method for their quantitative analysis [25–27]. The determination of oxygen ions in the molten salt can also be achieved by electrochemical titration with YSZE (yttria-stabilized zirconia electrode). Castrilejo et al. discussed the concentration changes of oxygen ion in $CaCl_2$-NaCl molten salt [28]. In addition, the square wave voltammetry was used to determine the relationship between current density and oxygen content in LiF-NaF-KF and LiF-NaF molten salts by Shen et al. [29] and Massot et al. [30], respectively. Thereby, three options for determining the oxygen content in molten salts can be concluded, namely: the best-fitting method, electrochemical titration technique (ETT), and square wave voltammetry (SWV).

Taking ETT as an example, in order to clarify the behavior of the impurity element (oxide) in molten salt, the electrochemical behavior of oxide ions has been investigated in an equimolar NaCl-KCl by electrochemical titration technique. The residual content of oxide in the molten NaCl-KCl was determined by using this in-situ method. The titrations of O^{2-} were applied by sodium oxide additions. The results obtained through electrochemical titration technique show that the oxide content in the melt and the free oxide ions (O^{2-}) detected content have a good linear relationship (**Figure 7**).

The equation of the fitted line is:

$$I_p = 5.71 \times 10^{-5} + 4.56 \times 10^{-6} x_{O}{}^{2-} \tag{13}$$

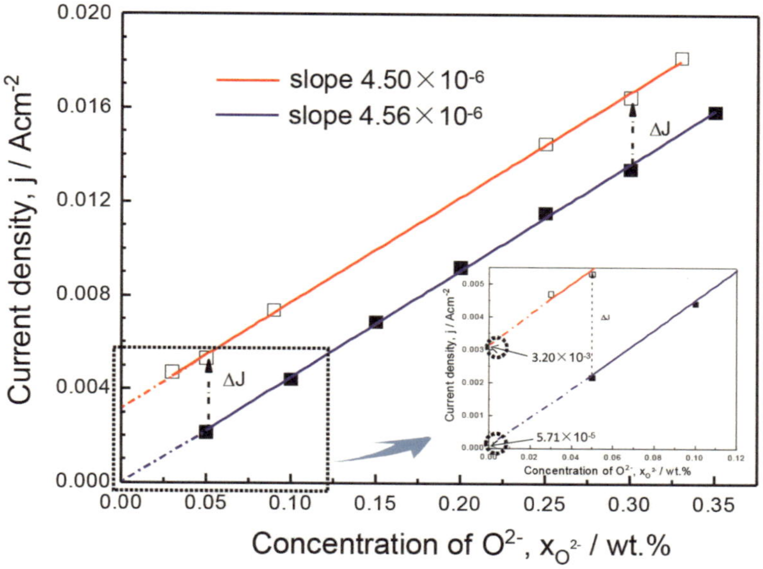

Figure 7.
Linear relationship of the peak current density versus oxide ion content before and after purified with HCl (reprinted from Ref. [27]).

The valve difference of slope can be explained as: the influence of O^{2-} decreased when its total concentration was on a high level. The line almost goes through the origin of the coordinates (the intercept equals to 5.71×10^{-5}), which means hardly any O^{2-} oxidation current can be detected with extremely low O^{2-} concentration dissolution in the NaCl-KCl melts. The current density differences (ΔJ) at a same concentration of O^{2-} are 0.0029 and 0.0027 Acm^{-2}, respectively. Thus, the residual concentration of O^{2-} can be estimated when plugging ΔJ into the Eq. (13), and the average value is 688 ppm.

3.4 Electro-refining of high-purity titanium in molten salt

3.4.1 Anodic polarization behavior in different molten salt structures

Oxidation occurs on the anode during electrolysis, and the anode material enters the molten salt in the form of ions. The electrochemical dissolution behavior is related to the electrolysis voltage, electrolyte composition, etc. [8, 31, 32]. Haarberg et al. investigated the electrochemical behavior of TiC_xO_y in NaCl-KCl and TiO_2 in Na_3AlF_6 molten salt. In the chloride molten salt, titanium ions are mainly dissolved in the form of Ti^{2+}, and the electrochemical precipitation of titanium ions is in the form of Ti^{3+} or Ti^{4+} in the molten salt of Na_3AlF_6.

It is of great significance for optimizing molten salt electrolyte and improving electrolysis efficiency. Thus, the changes of micro-element structure in molten salt, and the influence of electrochemical dissolution and deposition of titanium ions should be clarified.

3.4.2 Electrochemical parameters for high-purity titanium extraction

The reason for the low efficiency in electrolysis process is that, power consumption caused by the coordination of titanium ions and anions in the molten salt. The physical and chemical properties of the molten salt relate to the basis of electrolysis electrode process in molten salt. Thereby, the current efficiency of high-purity titanium electro-refining is related to the current density and the stable state of titanium ion in the molten salt. Concentrations of titanium ions and other parameters are also related [33–35]. Therefore, in order to achieve high-purity titanium with high-efficiency, the molten salt structure should be in-depth understood. It can be provided a guidance for improving the electrolysis efficiency and product quality.

4. Conclusions

Based on above analysis, comprehensive and systematic conclusions were done on the scientific issues to be clarified in the high-purity titanium molten salt electro-refining process. The influence of anion and cation in electrolyte on ionic equilibrium of titanium was discussed systematically by chemical analysis and electrochemical testing. Specifically, the type of microstructure of the molten salt is dissected in the molten salt mainly composed of chloride. The coordination mechanism between titanium ion and fluoride ion was also revealed. The mechanism of anodic oxidation and cathodic reduction kinetics in electrolytes with different micro-structures was clarified, which provide a guiding line for optimizing electrolyte. More importantly, on the basis of full theoretical and experimental research, it proposes parameters for high-purity titanium extraction in molten salt, and provides theoretical support for the establishment of efficient, low-pollution, and low-cost metallurgical processes.

Acknowledgements

The authors thank support from Collaborative Innovation Center of Henan Resources and Materials Industry, Zhengzhou University and Startup Research Fund of Zhengzhou University (No. 32210804). The authors are grateful to the National Natural Science Foundation of China (No. 51804277) and Key Projects of Henan Province Department of Education (No.19B450004).

Author details

Jianxun Song, Yusi Che*, Yongchun Shu and Jilin He
Henan Province Industrial Technology Research Institute of Resources and Materials, Zhengzhou University, Zhengzhou, China

*Address all correspondence to: cheys@zzu.edu.cn

References

[1] Li C, Song J, Li S, Li X, Shu Y, He J. An investigation on electrodeposition of titanium in molten LiCl-KCl. TMS, Materials Processing Fundamentals. 2019:193-202

[2] Pore VJ, Okura S, Suemori H. Process for deposition of titanium oxynitride for use in integrated circuit fabrication. 2016. US Patent 9523148

[3] Girginov A, Tzvetkoff T, Bojinov M. Electrodeposition of refractory metals (Ti, Zr, Nb, Ta) from molten salt electrolytes. Journal of Applied Electrochemistry. 1995;**25**:993-1003

[4] Wendt H, Reuhl K, Schwarz V. Cathodic deposition of refractory intermetallic compounds from FLiNaK melts I. Voltammetric investigation of Ti, Zr, B, TiB_2 and ZrB_2. Electrochimica Acta. 1992;**37**:237-244

[5] Song J, Xiao J, Zhu H. Electrochemical behavior of titanium ions in various molten alkali chlorides. Journal of the Electrochemical Society. 2017;**164**(12):E321-E325

[6] Kang M, Song J, Zhu H, Jiao S. Cathodic deposition process of the titanium in $CaCl_2$-$TiCl_2$ molten salt. Metallurgical and Materials Transactions B. 2015;**46**:162-168

[7] Martinez AM, Castrillejo Y, Barrado E, Haarberg GM, Picard G. A chemical and electrochemical study of titanium ions in the molten equimolar $CaCl_2$-NaCl mixture at 550°C. Journal of Electroanalytical Chemistry. 1998;**449**:67-80

[8] Haarberg GM, Kjos OS, Martinez AM, Osen KS, Skybakmoen E, Dring K. Electrochemical behaviour of dissolved titanium species in molten salts. ECS Transactions. 2010;**33**:167-173

[9] Lantelme F, Kuroda K, Barhoun A. Electrochemical and thermodynamic properties of titanium chloride solutions in various alkali chloride mixtures. Electrochimica Acta. 1998;**44**:421-431

[10] Jiao H, Wang J, Zhang L, Zhang K, Jiao S. Electrochemically depositing titanium (III) ions at liquid tin in a NaCl-KCl melt. RSC Advances. 2015;**5**: 62235-62240

[11] Ning X, Asheim H, Ren H, Jiao S. Preparation of titanium deposit in chloride melts. Metallurgical and Materials Transactions B. 2011;**42**: 1181-1187

[12] Haarberg GM, Rolland W, Sterten A, Thonstad J. Electrodeposition of titanium from chloride melts. Journal of Applied Electrochemistry. 1993;**23**: 217-224

[13] Song Y, Jiao S, Hu L, Guo Z. The cathodic behavior of Ti(III) ion in a NaCl-2CsCl melt. Metallurgical and Materials Transactions B. 2016;**47**(1): 804-810

[14] Song J, Wang Q, Kang M, Jiao S, Zhu H. The equilibrium between $TiCl_2$, $TiCl_3$ and metallic titanium in molten binary mixtures of LiCl. Electrochemistry. 2014;**82**:1047-1051

[15] Lantelme F, Salmi A. Electrochemistry of titanium in NaCl-KCl mixtures and influence of dissolved fluoride ions. Journal of the Electrochemical Society. 1995;**142**: 3451-3456

[16] Polyakova LP, Taxil P, Polyakov EG. Electrochemical behaviour and codeposition of titanium and niobium in chloride-fluoride melts. Journal of Alloys and Compounds. 2003;**359**: 244-255

[17] Robin A, Lepinay J. Determination of the apparent standard potential of the

Ti/Ti(III) system in the LiF-NaF-KF eutectic using voltammetry chronopotentiometry and open circuit potentiometry. Electrochimica Acta. 1991;**36**:1009-1012

[18] Song J, Wang Q, Zhu X, Hou J, Jiao S, Zhu H. The influence of fluoride anion on the equilibrium between titanium ions and electrodeposition of titanium in molten fluoride-chloride salt. Materials Transactions. 2014;**55**(8): 1299-1303

[19] Song J, Wang Q, Jiao S, Zhu H. The influence of fluoride ions on the equilibrium between titanium ions and titanium metal in fused alkali chloride melts. Faraday Discussions. 2016;**190**: 421-432

[20] Song J, Huang X, Wu J, Zhang X. Electrochemical behaviors of Ti(III) in molten NaCl-KCl under various contents of fluoride. Electrochimica Acta. 2017;**256**:252-258

[21] Sekimoto H, Nose Y, Uda T, Uehara A, Yamana H, Sugimura H. Revaluation of equilibrium quotient between titanium ions and metallic titanium in NaCl-KCl equimolar molten salt. Journal of Alloys and Compounds. 2011;**509**:5477-5482

[22] Wang Q, Song J, Hu G, Zhu X, Hou J, Jiao S, et al. The equilibrium between $TiCl_2$, $TiCl_3$ and titanium metal in NaCl-KCl equimolar molten salt. Metallurgical and Materials Transactions B. 2013;**44B**:906-913

[23] Zhu X, Wang Q, Song J, Hou J, Jiao S, Zhu H. The equilibrium between metallic titanium and titanium ions in LiCl-KCl melts. Journal of Alloys and Compounds. 2014;**587**:349-353

[24] Barner J, Noye P, Barhoun A, Lantelme F. Influence of oxide and alloy formation on the electrochemistry of Ti deposition from the NaCl-KCl-NaF-K_2TiF_6 melt reduced by metallic Ti. Journal of the Electrochemical Society. 2005;**152**(1):C20-C26

[25] Bright FH, Wurm JG. Some new fluoride complexes of trivalent titanium. Canadian Journal of Chemistry. 1958;**36**:615-622

[26] Song J, Mukherjee A. Influence of F^- on the electrochemical properties of titanium ions and Al-Ti alloy electrodeposition in molten $AlCl_3$-NaCl. RSC Advances. 2016;**6**:82049-82056

[27] Song J, Huang X, Fan Y, Yi J, Shu Y, He J. In-situ monitoring of O^{2-}concentration in molten NaCl-KCl at 750°C. Journal of the Electrochemical Society. 2018;**165**(5):E245-E249

[28] Castrillejo Y, Martinez A, Haarberg GM, Borresen B, Osen K, Tunold R. Oxoacidity reactions in equimolar molten $CaCl_2$-NaCl mixture at 575°C. Electrochimica Acta. 1997;**42**: 1489-1494

[29] Shen M, Peng H, Ge M, Zuo Y, Xie L. Use of square wave voltammeter for online monitoring of O^{2-} concentration in molten fluorides at 600°C. Journal of Electroanalytical Chemistry. 2015;**748**:34-39

[30] Massot L, Cassayre L, Chamelot P, Taxil P. On the use of electrochemical techniques to monitor free oxide content in molten fluoride media. Journal of Electroanalytical Chemistry. 2007;**606**:17-23

[31] Wang Q, Song J, Wu J, Jiao S, Hou J, Zhu H. A new consumable anode material of titanium oxycarbonitride for the USTB titanium process. Physical Chemistry Chemical Physics. 2014;**16**: 8086-8091

[32] Jiao S, Ning X, Huang K, Zhu H. Electrochemical dissolution behavior of conductive TiC_xO_{1-x} solid solutions. Pure and Applied Chemistry. 2010; **82**(8):1691-1699

[33] Wu J, Song J, Zhu H, Shu Y, He J. Equilibrium between metallic titanium and titanium ions in $MgCl_2LiCl$ molten salt. Materials Transactions. 2019;**60**(3): 374-378

[34] Song J, Wang Q, Hu G, Zhu X, Jiao S, Zhu H. The equilibrium between titanium ions and the electro-refining of high-purity titanium in a NaCl-KCl melt. International Journal of Minerals Metallurgy and Materials. 2014;**21**(7): 660-665

[35] Song J, Hu G, Wang Q, Jiao S, Zhu H. Preparation of high purity titanium by electro-refining in NaCl-KCl Melt. In: The 4th Asian Conference on Molten Salts and Ionic Liquids, 44th Symposium on Molten Salt Chemistry; Sendai, Japan; 2012. pp. 120-126

Chapter 6

Ionothermal Synthesis of Metal-Organic Framework

Hyun-Chang Oh, Sukwoo Jung, Il-Ju Ko and Eun-Young Choi

Abstract

Ionothermal synthesis employs ionic liquids for synthesis of metal organic frameworks (MOFs) as solvent and template. The cations and anions of ionic liquids may be finely adjusted to produce a great variety of reaction environments and thus frameworks. Organisation of the structures synthesised from related ionic liquid combinations give rise to provocative chemical trends that may be used to predict future outcomes. Further analysis of their structures is possible by reducing the complex framework to its underlying topology, which by itself brings more precision to prediction. Through reduction, many seemingly different, but related classes of structures may be merged into larger groups and provide better understanding of the nanoscopic structures and synthesis conditions that gave rise to them. Ionothermal synthesis has promised to enable us to effectively plan the synthesis ahead for a given purpose. However, for its promise to be kept, several difficult limitations must be overcome, including the inseparable cations from the solvent that reside in the framework pore.

Keywords: ionothermal synthesis, metal organic frameworks, imidazolium-based ionic liquids, chemical trend analysis, structure simplification

1. Introduction

Three things are necessary to consider in the preparation of metal organic frameworks (MOFs): the metal, organic ligand, and solvent. Often neglected is the influence exerted by the solvent on the eventual framework, unlike the metal and the ligand that the structure always constitutes of. Varying the key characteristics of the solvent, such as hydrophilicity, is often the deciding factor in the reaction yield and the nature of the final compound [1]. Until 2002, when Jin et al. first used ionic liquids to synthesise metal organic framework [2], the list of solvents in inorganic synthesis was limited to few organic solvents and water [3]. This new synthesis method received growing attention in the field of MOFs to open a new realm of novel structures and provocative findings regarding the very nature of nanoscale synthesis. One aspect of ionothermal synthesis that contributed to its attention must have been its simplicity; the overall process comprises no more than mixing the metal salt and the organic ligand with the ionic solvent and incubating at a high temperature for long enough time. Unfortunately, however, the growth seems to have ceded in the recent years as shown in **Figure 1**. Given its distinctive potentials, this chapter is dedicated to introduce the field and draw more efforts for the full realisation of what the methodology dare to have promised. Before we move to the discussion of ionothermal synthesis and its potentials, the chemistry of ionic liquids

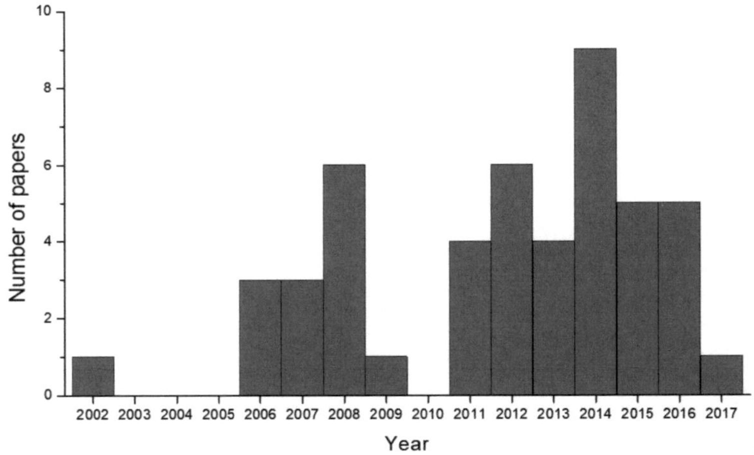

Figure 1.
Number of papers published annually under the topic of the ionothermal synthesis of MOFs since its first report.

must be first visited since it is this distinctive nature that lies behind all positive aspects, and limitations too, of ionothermal synthesis.

Ionic liquids are simply salts in the liquid state as opposed to the liquids typically used as solvents, [4] which are predominantly comprised of electrically neutral molecules. While most salts may be brought to their liquid states by heating, the term 'ionic liquids' is exclusively used for those that stay fluid around or below 100°C to distinguish them from the older phrase 'molten salts' [5]. One reason behind the attention that ionothermal methodology receives may be directly induced from the term 'ionic liquid' itself. The liquids are held by ionic interactions that far outcompetes the most intermolecular interactions in other solvents, including the renowned hydrogen bonds in water. Such strong interaction is responsible for their low vapour pressure [6], which could resolve the safety and environmental concerns associated with conventional organic solvents [5]. Such characteristics function as the exact same advantages in synthesis of MOFs. Nevertheless, the synthesis has greater potential in which the reaction environment can be finely tuned by modifying the solvent ions [7]. There are only several hundreds of molecular solvents, whereas a million binary combinations and a million of millions of ternary combinations possible for ionic liquids [5], hence their nickname 'designer solvent [8]'. Efforts in the field need to be focused not only on collecting outcomes from as many combinations as possible, but also, more importantly, on comprehending the laws of chemistry lying behind the trend observable in those data. This shall, as more than enough possible combinations await, ultimately enable designing the product for a given purpose, rather than vice versa.

This chapter will focus on showing the potentials of ionothermal synthesis by presenting a set of related syntheses in an organised manner. A series of such ionic liquids (RMI-X) may be prepared with 1-alkyl-3-methylimidazolium (RMI) and halide ions(X) [9]. This series of solvents exhibit finest tunability, in addition to their stability, with its variable length of alkyl side chain of the cation and the anion species variable along the halide column, which places them among the most extensively studied solvents for ionothermal synthesis of MOFs. To confirm their dominance in structure reports and the focus of our discussion on them, investigations have been made about the number of MOFs synthesised with several popular ionic liquids. The scope of our search—the list of cations and anions comprising the most common ionic liquids—has been illustrated in **Figure 2**. According to Cambridge Structural Database (CSD), it was shown that much of the reported MOFs is synthesised from

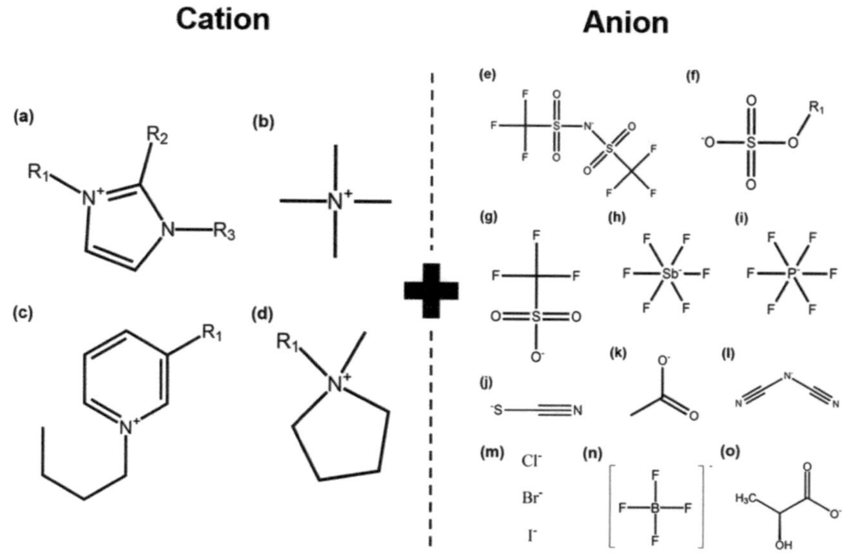

Figure 2.
Common cations and anions that composes an ionic liquid. Different combination of cations and anions results in various characteristics and gives rise to the diversity of topology and structure in synthesised MOFs. (a) Imidazolium (R_1 = methyl, ethyl, butyl; R_2 = hydrogen, methyl; R_3 = methyl, ethyl, propyl, butyl, pentyl, hexyl, octyl, benzyl, allyl, vinyl); (b) Ammonium; (c) Pyridinium (R_1 = hydrogen, methyl); (d) Pyrrolidinium (R_1 = ethyl, propyl, butyl, hexyl); (e) Bis((trifluoromethyl)sulfonyl)amide; (f) Alkyl sulphate (R_1 = methyl, ethyl); (g) Trifluoromethanesulfonate; (h) Hexafluorostibate(V); (i) Hexafluorophosphate(V); (j) Thiocyanate; (k) Acetate; (l) Bis(cyanide)amide; (m) Halide ion; (n) Tetrafluoroborate; (o) L-lactate.

ionic liquids that contains imidazolium and halide ions. There are no MOF crystal containing pyridium cation, and only few crystals synthesised from tetramethyl ammonium is reported as MOF. Synthesis using pyrrolidinium cation shows about 100 crystals, which corresponds to co-crystal form, showing that no crystal exhibits MOFs including pyrrolidinium cation. Extensiveness of data is the foundation of all successful discussions. With the extensiveness of RMI-X now taken for granted, structures synthesised from conditions with piecemeal differences, namely the length of the side chain of the cation, halide ions, and core metal atom of the structure were analysed to explore the effect on the final product arising from such variations.

2. Gradual difference in the solvent brings about gradual difference in the product

An important characteristic of ionothermal synthesis is that the characteristics of the solvents may be gradually varied and investigate the difference induced in the final product. While the solvent can be substituted with a complete different class of cations or anions to provide a completely reshaped environment, more minor changes can be made to the ions so that the change is gradual and quantifiable. Changing the length of the alkyl side chains attached to imidazolium cations, or changing the anions within the halide column to gradually change the size of the solvent ions is one example that will be mainly discussed in the chapter. This way, we may grasp a better understanding of the relation between the beginning and the end of this nanoscopic synthesis. Actually, organic solvents hold the exact same advantage, seeing that even the size variation of imidazolium cations is actually an organic one. However, in ionic liquids, this variation is expanded to a two-dimensional table for binary combinations, and possibly to even four-dimensional construct for ternary combinations, which can provide more organised data is obvious. A better understanding of the nature is a foundation for a better utilisation of chemistry for many types of benefits. This section will guide you to the exploration that searches for new meaningful correlations in the sea of ionothermally prepared materials as the size of the solvent.

2.1 Correlation between the solvent and the product is often very simple

The correlation between the solvent and the product is perhaps the easiest to perceive in the system of frameworks synthesised with nickel and 1,3,5-benzenetricarboxylic acid(BTC). In **Table 1**, the structure shifts from the A-topology to the B-topology as we move down to the table and increase the cation size [1]. The shift occurs at smaller cations when we move right to the table to increase the anion size. It appears the size of the solvent, the cation and the anion considered together, is the key factor in determining the topology between A and B. From just this trend alone, it may be inferred that the B topology has a larger pore size, that is more empty space in the framework, than the A topology, which complies with the framework analyses by X-ray diffractions. There are certainly more reasons to this and we will come back to this later in the chapter, but for now, it is enough to just appreciate the simplicity of trend analysis.

The shift in the size of the ionic liquid exerted strong enough a pressure to give rise to two totally different topologies, but sometimes, the shift may be minor. In manganese-BTC system presented in **Table 2**, all three combinations in the [EMI] row gave rise to the exact same structure, α1 [10]. However, in the [PMI] column, only chloride and bromide gave rise to α2, and iodide to a slightly different α3. It is predicted that [EMI] cation is too small to induce a structure transition to occur in the row, but [PMI] is big enough to do so. Even though all the reported cases

$Ni(OAc)_2{\cdot}H_3BTC$	Cl	Br	I
[EMI]	A1	A1	A1
[PMI]	A2	A2	B1
[BMI]	A3	B2	B2
[PEMI]	B3	B3	B3

Structures sharing a topological identity were labelled under the same alphabet, while the numbers denote minor difference among them. Each labels denote, $[RMI]_2[Ni_3(BTC)_2(OAc)_2]$ (RMI = EMI for A1, PMI for A2, BMI for A3), $[RMI]_2[Ni_3(HBTC)_4(H_2O)_2]$ (RMI = BMI for B2, PEMI for B3). A1, A2, A3, B1, B2 reported in [1], and B3 in [7].

Table 1.
Organisation of structures in Nickel-BTC system on the length of the alkyl side chain of the cation and the halide anion.

$Mn(OAc)_2{\cdot}H_3BTC$	Cl	Br	I
[EMI]	α1	α2	α1
[PMI]	α2	α2	α3
[BMI]	—	—	—
[PEMI]	—	—	—

Each labels denote, α1-[EMI][Mn(BTC)], α2-[PMI][Mn(BTC)], α2-[PMI][Mn(BTC)]. Combinations that have not been reported were left blank. All entries were reported in [10].

Table 2.
*A table for the system of framework synthesised with manganese-BTC system arranged similarly to **Table 1**.*

in the system belong to the same topology class, but when the smaller differences were accounted, the table again shows a similar stair-shaped pattern that may be explained using the exact same argument.

$Cd(OAc)_2$·H_3BTC	Cl	Br	I
[EMI]	a1 [12]	b1 [12, 13]	b1 [12]
[PMI]	b2 [12]	b2 [12]	b2 [12]
[BMI]	—	—	—

Each labels denote, a1-[EMI][Cd2(BTC)Cl2] for a1, [RMI][Cd(BTC)] (EMI for b1, PMI for b2). Combinations that have not been reported were left blank. Reference in literature to which the entries may be corresponded to has been indicated next to the entries in the table.

Table 3.
A table for the system of framework synthesised with cadmium-BTC system arranged similarly to ***Table 1****.*

M	Formula	Nuclear #	CCDC code	Void with cation*	Void without cation*
Co	$[EMI]_2[Co_3(BDC)_3Cl_2]$	Tri-nuclear	TACHUD [14]	0% 0 Å^3 /2088.3 Å^3	39.6% 827.3 Å^3 /2088.3 Å^3
	$[PMI]_2[Co_3(BDC)_3Cl_2]$	Tri-nuclear	TACJAL [14]	0% 0 Å^3 /2106.8 Å^3	39.8% 839.6 Å^3 /2106.8 Å^3
	$[BMI]_2[Co_3(BDC)_3Cl_2]$	Tri-nuclear	TACJEP [14]	0% 0 Å^3 /2082.1 Å^3	40.0% 832.7 Å^3 /2082.1 Å^3
	$[AMI]_2[Co_3(BDC)_3Cl_2]$	Tri-nuclear	TACJIT [14]	0% 0 Å^3 /2314.1 Å^3	46.8% 1083.9 Å^3 /2314.1 Å^3
	$[EMI]_2[Co_3(BDC)_3Br_2]$	Tri-nuclear	JOQXOE [15]	1.7% 35.2 Å^3 /2103.5 Å^3	38.9% 817.7 Å^3 /2103.5 Å^3
	$[PMI]_2[Co_3(BDC)_3Br_2]$	Tri-nuclear	JOQXUK [15]	0.6% 13.0 Å^3 /2123.4Å^3	40.0% 848.5 Å^3 /2123.4 Å^3
	$[BMI]_2[Co_3(BDC)_3Br_2]$	Tri-nuclear	JOQPUC [15]	0% 0 Å^3 /2172.6 Å^3	40.4% 877.1 Å^3 /2172.6 Å^3
	$[AMI]_2[Co_3(BDC)_4]$	Tri-nuclear	JOQQAJ [15]	17.0% 1001.9 Å^3 /5898.3 Å^3	52.2% 3081.8 Å^3 /5898.3 Å^3
	$[EMI]_2[Co_3(BDC)_3I_2]$	Tri-nuclear	TACJOZ [14]	1.4% 31.0 Å^3 /2184.2 Å^3	40.3% 879.5 Å^3 /2184.2 Å^3
	$[PMI]_2[Co_3(BDC)_3I_2]$	Tri-nuclear	TACJUF [14]	0.9% 20.9 Å^3 /2201.7Å^3	41.5% 914.8 Å^3 /2201.7 Å^3
	$[BMI]_2[Co_3(BDC)_3I_2]$	Tri-nuclear	TACKAM [14]	0% 0 Å^3 /2267.4Å^3	39.7% 901.0 Å^3 /2267.4 Å^3
	$[AMI]_2[Co_3(BDC)_3I_2]$	Tri-nuclear	TACKEQ [14]	0% 0 Å^3 /2275.1 Å^3	44.7% 1017.9 Å^3 /2275.1 Å^3

M	Formula	Nuclear #	CCDC code	Void with cation*	Void without cation*
Zn	$[EMI]_2[Zn_3(BDC)_3Cl_2]$	Tri-nuclear	SIVQEV [16]	0% 0 Å^3 /2085.2 Å^3	39.6% 826.7 Å^3 /2085.2 Å^3
	$[PMI]_2[Zn_3(BDC)_3Cl_2]$	Tri-nuclear	QUGVAR [17]	0% 0 Å^3 /2097.0Å^3	39.7% 833.2 Å^3 /2097.0 Å^3
	$Zn(BDC)(H_2O)$	Poly-nuclear	IFABIA03 [17]	0% 0 Å^3 /826.8 Å^3	0% 0 Å^3 /826.8Å^3
	$[BMI]_2[Zn_3(BDC)_3Cl_2]$	Tri-nuclear	SIVCAD [16]	0% 0 Å^3 /2167.9 Å^3	41.7% 903.4 Å^3 /2167.9 Å^3
	$[EMI]_2[Zn_3(BDC)_3Br_2]$	Tri-nuclear	QUGVIZ [17]	0.8% 16.5 Å^3 /2115.0 Å^3	39.9% 844.0 Å^3 /2115.0 Å^3
	$[PMI]_2[Zn_3(BDC)_3Br_2]$	Tri-nuclear	QUGVOF [17]	0.6% 11.9 Å^3 /2139.4 Å^3	40.6% 868.6 Å^3 /2139.4 Å^3
	$[BMI]_2[Zn_3(BDC)_3Br_2]$	Tri-nuclear	QUGVUL [17]	0% 0 Å^3 /2189.2 Å^3	42.1% 921.1 Å^3 /2189.2 Å^3
	$[EMI]_2[Zn_3(BDC)_3I_2]$	Tri-nuclear	QUGWEW [17]	0.6% 12.1 Å^3 /2169.2 Å^3	40.0% 868.7 Å^3 /2169.2 Å^3
	$[PMI]_2[Zn_3(BDC)_3I_2]$	Tri-nuclear	QUGWIA [17]	0% 0 Å^3 /2182.0 Å^3	40.4% 882.6 Å^3 /2182.0 Å^3
	$[BMI]_2[Zn_3(BDC)_3I_2]$	Tri-nuclear	QUGWOG [17]	0% 0 Å^3 /2253.3 Å^3	39.7% 894.2 Å^3 /2253.3 Å^3
	$[AMI]_2[Zn_3(BDC)_3I_2]$	Tri-nuclear	QUGWUM [17]	0% 0 Å^3 /2293.1 Å^3	40.7% 933.1 Å^3 /2293.1 Å^3
Eu	$[EMI][Eu_2(BDC)_3Cl]$	Di-nuclear	IXISOZ02 [18]	0%, 0 Å^3 /2339.26 Å^3	30.5% 714.34 Å^3 /2339.3 Å^3
	$[PMI][Eu_2(BDC)_3Cl]$	Di-nuclear	IXITOA02 [18]	4.6% 157.3 Å^3 /3429.0 Å^3	30.8% 1055.9 Å^3 /3429.0 Å^3
	$[BMI][Eu_2(BDC)_3Cl]$	Di-nuclear	IXITIU02 [18]	0.2% 14.7 Å^3 /6850.9 Å^3	29.8% 2038.9 Å^3 /6850.9 Å^3
	$[EMI][Eu_2(BDC)_3(H_2BDC)Cl_2]$	Poly-nuclear	YIXFOC03 [19]	0% 0 Å^3 /3153.1Å^3	21.4% 673.7 Å^3 /3153.1 Å^3
	$Eu(BDC)(CO_2)$	Poly-nuclear	LARYEK03 [20]	0% 0 Å^3 /929.6 Å^3	0% 0 Å^3 /929.6 Å^3
	$Eu_3(BDC)_4Cl(H_2O)_6$	Poly-nuclear Mono-nuclear	IXITEQ02 [18]	0% 0 Å^3 /3605.0 Å^3	0% 0 Å^3 /3605.0 Å^3
	$Eu(BDC)Cl(H_2O)$	Poly-nuclear	IXISIT02 [18]	0% 0 Å^3 /975.0 Å^3	0% 0 Å^3 /975.0 Å^3

M	Formula	Nuclear #	CCDC code	Void with cation*	Void without cation*
In	$[EMI]_2[In_2(BDC)_3Br_2]$	Mono-nuclear	SABJOX [21]	0% 0 Å^3 /2017.0 Å^3	36.0% 725.1 Å^3 /2017.0 Å^3
	$[PMI]_2[In_2(BDC)_3Br_2]$	Mono-nuclear	SABJIR [21]	0% 0 Å^3 /2073.5 Å^3	36.0% 746.0 Å^3 /2073.5 Å^3
Dy	$[EMI][Dy_3(BDC)_5]$	Poly-nuclear	RINTUF [22]	2.8% 134.8 Å^3 /4840.8 Å^3	21.2% 1027.3 Å^3 /4840.8 Å^3
Tb	$[EMI][Tb_2(\mu_2\text{-}Cl)(BDC)_3]$	Poly-nuclear	YIXFUI [19]	19.7% 620.8 Å^3 /3152.5 Å^3	19.7% 620.8 Å^3 /3152.5 Å^3
Sm	$[EMI]_2[Sm_2(BDC)_3(H_2\text{-}BDC)Cl_2]$	Di-nuclear	YIXFIW [19]	0% 0 Å^3 /2333.3 Å^3	30.4% 709.1 Å^3 /2333.3 Å^3
Cd	$[BMI]_2[Cd(BDC)_3Br_2]$	Tri-nuclear	QETDAV [23]	0% 0 Å^3 /2353.8 Å^3	40.4% 950.6 Å^3 /2353.8 Å^3
Cu	$[EEIM][NaCu(BDC)_2]$	Poly-nuclear	VOBRUB [24]	0% 0 Å^3 /2200.5 Å^3	31.3% 688.2 Å^3 /2200.5 Å^3

**Probe radius of 1.2 Å and grid spacing of 0.7 Å was taken for the calculation using the contact surface.*

Table 4.
Chemical formulas of structures arising from imidazolium-based MOFs with BDC are presented with their nuclear type, CCDC reference code, void volume with and without the residing cation, and the reference in literature that each structure was reported.

2.2 Ionic liquids function both as a solvent and template

Similar trends may also be found in other metals, despite less well-pronounced than nickel. The similarity may not be noticed at first glance, but it is the same stair-shaped pattern to nickel system. The topology shift just takes place with smaller ionic species. Again, increase in size of the ionic solvent has changed the preferred topology to another class with a larger pore volume to incorporate the ions. As some readers might have noticed by now, here is a good point to introduce another interesting aspect of ionothermal synthesis; ionic liquids function not only as solvents, but also as a template that physically exerts a pressure to determine the final topology by residing in the framework [11] (**Table 3**).

2.3 Many reported syntheses are yet to fit into an organised system

In theory, many choices of organic linkers available in the field of chemistry may add to the large number of ionic combinations to create nearly infinite possible cases, but it takes time for a theory to become reality. While many valuable efforts are being made to contribute, those with 1,3,5-bezenetricarboxylic acid(BTC) and 1,4-benzenedicarboxylic acid(BDC) have done its part particularly extensively and the reported structures were organised in **Tables 4** and **5**. **Tables 4** and **5** are great to appreciate the variety of ionothermally prepared MOFs, plus for searching purposes, but give hardly any information on the chemical reaction that brought about the structures. In order to get a closer grasp on how ionothermal synthesis produced this variety, they must be organised into systems of related syntheses. However, many cases in both tables are rather discrete. Efforts need to be made, starting from what have been reported, to expand the literature by applying the ionic liquid to gradual variations.

M	Formula	Nuclear #	CCDC	Void with cation*	Void without cation*
Ni	$[EMI]_2[Ni_3(BTC)_2(OAc)_2]$	Tri-nuclear	VEMSUC [25]	0% 0 Å^3 /3704 Å^3	36.5% 1350.8 Å^3 /3704 Å^3
	$[PMI]_2[Ni_3(BTC)_2(OAc)_2]$	Tri-nuclear	XUJPIC [1]	0.9% 26.4 Å^3 /2780.4 Å^3	38.7% 1076.8 Å^3 /2780.4 Å^3
	$[BMI]_2[Ni_3(BTC)_3(OAc)_2]$	Tri-nuclear	XUJPOI [1]	0% 0 Å^3 /3802.9 Å^3	0% 0 Å^3 /3802.9 Å^3
	$[PMI]_2[Ni_3(HBTC)_4(H_2O)]$	Tri-nuclear	VEMSUC01 [1]	0% 0 Å^3 /3712.3 Å^3	36.7% 1363.0 Å^3 /3712.3 Å^3
	$[BMI]_2[Ni_3(HBTC)_4(H_2O)]$	Tri-nuclear	XUJQAV [1]	0% 0 Å^3 /3806.2 Å^3	0% 0 Å^3 /3806.2 Å^3
	$[BMI]_2[Ni(HBTC)_2(H_2O)_2]$	Tri-nuclear	NUNNUH [26]	3.4% 132.7 Å^3 /3944.7 Å^3	53.6% 2113.7 Å^3 /3944.7 Å^3
	$[BMI]_2[Ni(HBTC)_2(H_2O)_2]$ chiral	Tri-nuclear	NUNPAP [26]	3.2% 124.9 Å^3 /3953.2 Å^3	53.6% 2120.3 Å^3 /3953.2 Å^3
	$[AMI][Ni_3(BTC)_2(OAc)(MI)_3]$	Mono-nuclear Di-nuclear	EGOYUV [27]	0% 0 Å^3 /4592.0 Å^3	22.4% 1029.7 Å^3 /4592.0 Å^3
Co	$[EMI][Co_2(HBTC)_2(4,4'\text{-bpy})_3]Br$	Di-nuclear	YODYUM [28]	0% 0Å^3 /5060.8 Å^3	13.6% 688.9 Å^3 /5060.8 Å^3
	$[EMI][Co(HBTC)(4,4'\text{-bpy})_2](4,4'\text{-bpy})Br$	Mono-nuclear	YODZAT** [28]	0% 0 Å^3 /4290.3 Å^3	17.8% 762.0 Å^3 /4290.3 Å^3
	$[EMI][Co(BTC)(H\text{-}Im)]$	Mono-nuclear	YODZEX [28]	0% 0 Å^3 /933.9 Å^3	30. 1% 280.63 Å^3 /933.9 Å^3
	$[EMI]_2[Co(BTC)_2(H_2TED)]$	Mono-nuclear	YODZIB [28]	0% 0 Å^3 /7246.2 Å^3	30.7% 2221.6 Å^3 /7246.2 Å^3
	$[EMI]_2[Co_3(BTC)_2(OAc)_2]$	Tri-nuclear	VEMTAJ [25]	38.6% 1466.9 Å^3 /3797.9 Å^3	38.6% 1466.9 Å^3 /3797.9 Å^3
	$[EMI][Co(BTC)]$	Di-nuclear	CIPLIX [29]	0% 0 Å^3 /3018.2 Å^3	26.0% 784.2 Å^3 /3018.2 Å^3
	$[EMI][Co_2(BTC)_4H_7(2,2'\text{-bpy})_2]$	Mono-nuclear	CIPLOD [29]	16.1% 483.0 Å^3 /3002.0 Å^3	16.1% 483.0 Å^3 /3002.0 Å^3
	$[PMI]_2[Co_2(BTC)_2(H_2O)_2]$	Di-nuclear	XAPSIS [30]	0% 0 Å^3 /1850.4 Å^3	51.1% 945.9 Å^3 /1850.4 Å^3
	$[BMI]_2[Co_2(BTC)_2(H_2O)_2]$	Di-nuclear	[31]	0% 0 Å^3 /1840.5 Å^3	51.8% 952.94 Å^3 /1840.5 Å^3
	$[EMI]_2[In_2Co(OH)_2(BTC)_2Br_2]$	Tri-nuclear (In + Co)	VUVZAP [32]	0% 0 Å^3 /1884.5 Å^3	36.6% 690.4 Å^3 /1884.5 Å^3

M	Formula	Nuclear #	CCDC	Void with cation*	Void without cation*
Zn	[BMI][Zn_2(BTC)(OH)I]	Tetra-nuclear	VIWZOR [33]	4.7% 106.8 Å^3 /2260.6 Å^3	49.4% 1117.8 Å^3 /2260.6 Å^3
	$[BMI]_2[Zn_4(BTC)_3(OH)(H_2O)_3]$	Tri-nuclear	MIQLEE [34]	4.8% 228.0 Å^3 /4739.5 Å^3	38.4% 1819.6 Å^3 /4739.5 Å^3
	[AMI][Zn_2(BTC)(OH)Br]	Tetra-nuclear	MIQLII [34]	0% 0 Å^3 /2229.3 Å^3	51.0% 1137.0 Å^3 /2229.3 Å^3
	[EMI][Zn(BTC)]	Di-nuclear	MIQKUT [34]	0% 0 Å^3 /3077.8 Å^3	38.2% 1175.6 Å^3 /3077.8 Å^3
	[PMI][Zn(BTC)]	Mono-nuclear	MIQLAA [34]	0% 0 Å^3 /3353.5 Å^3	43.9% 1472.9 Å^3 /3353.5 Å^3
	$[Zn_3(BTC)_2(H_2O)_2]\cdot 2H_2O$	Poly-nuclear	MISRUC [34]	0% 0 Å^3 /1985.1 Å^3	0% 0 Å^3 /1985.1 Å^3
	$[Zn_4(BTC)_2(\mu_4\text{-}O)(H_2O)_2]$	Tetra-nuclear	XOHLEM [35]	37.7% 1293.1 Å^3 /3431.7 Å^3	37.7% 1293.1 Å^3 /3431.7 Å^3
	[BMI][Zn(BTC)]	Di-nuclear	FUTZEB [36]	0% 0 Å^3 /1678.8 Å^3	46.8% 786.5 Å^3 /1678.8 Å^3
Cd	[EMI][$Cd_2(BTC)Cl_2$]	Poly-nuclear	SIZGEO [12]	0% 0 Å^3 /924.3 Å^3	34.0% 314.7 Å^3 /924.3 Å^3
	[EMI][Cd(BTC)]	Di-nuclear	NEHMET [13]	0% 0 Å^3 /3181.9 Å^3	43.4% 1381.4 Å^3 /3181.9 Å^3
	[PMI][Cd(BTC)]	Di-nuclear	SIZGIS [12]	0% 0 Å^3 /3296.9 Å^3	43.2% 1425.3 Å^3 /3296.9 Å^3
Mn	[EMI][Mn(BTC)]	Di-nuclear	WEYQAU [10]	0% 0 Å^3 /3051.7 Å^3	38.7% 1181.6 Å^3 /3051.7 Å^3
	[PMI][Mn(BTC)]	Di-nuclear	WEYQEY [10]	0% 0 Å^3 /3229.3 Å^3	42.8% 1382.5 Å^3 /3229.3 Å^3
	[PMI][Mn(BTC)]	Di-nuclear	WEYPUN [10]	0% 0 Å^3 /3172.0 Å^3	45.8% 1451.3 Å^3 /3172.0 Å^3

**Calculation was done with the same setting. Calculation for the entry 'void without cations' has been conducted by removing only the cations from the channel.*
***In the entry with reference code YODZAT, not only ionic cations but also 4,4'-bpy ligands were found in the channel.*

Table 5.
Structures arising from imidazolium-based MOFs with BTC presented in a similar manner to ***Table 1****.*

3. The correlation between the reaction solvent and the product

In the previous section, we have explored through the chemical trend observable in nickel, manganese, and cadmium-BTC systems. Some basic explanations have been provided by relating the size of the ionic species to the pore size of each structure. However, many questions still remain to be answered. For example, why does the topology shift has to occur right there, not anywhere else? If even larger cations were used, will the topology remain unchanged or will a new one appear? In order to answer this question, we have to get a deeper knowledge about the structures and the ions of the solvent. Qualitative analyses were simple, but it becomes necessary to add numbers to our logics to advance further.

3.1 Trends to predict future outcomes

In the nickel-BTC system we first examined, increasing the cation size caused the topology to shift from the more condensed A-class to the B-class with a larger cavity volume. In **Table 1**, data have been provided from [EMI] to [PEMI] only. Suppose we are interested in the results for a new [HMI] column; the cation alkyl side chain increased by a carbon. The simplest way to predict the outcome is to analyse the cavity of the B-class topology and test if it is capable of housing the elongated [HMI] cations. In addition, taking a deeper look into the A-class topology might also strengthen our previous reasoning.

In the A-class topology, we can see that the cavities are not so well connected with each other [1, 7]. On the other hand, the B-class topology has cavities connected to their neighbours to create a linear channel-shaped pore as illustrated in **Figure 3**. The maximum length of the cation that can fit into the A-class topology is limited, but it is not so in the B-class topology. After looking at how the structures actually look like, we can now more confidently say that elongating the cation by a carbon is unlikely to exert enough pressure on the framework to give rise to a new topology.

The next step is to test whether the predictions are correct. When a completely new material is synthesised and its structure is to be determined, advanced tools like single-crystal X-ray diffractions must be used to resolve all positions of the atoms in the unit cell. However, when we have a reference material with a known structure, simpler techniques like powder X-ray diffractions (PXRD) are enough to tell whether the new material has the same structure to the reference. Therefore, the common step is to obtain the PXRD graph and compare it to the graph of some suspected structures. It is only when the new graph is different enough, the new material is subjected to complete structure analysis. Since our expectation was that the entries for the [HMI] column will have the same topology to the [BMI] column, we took the PXRD data for two and compared as in **Figure 4**. To assure the topological identity, several different combinations have been selected. All major peaks occur at the same angles, and our prediction by extrapolation has been proven valid [7]. We have now seen that correlations are valuable in that they may be used to predict future outcomes. This is the moment where prediction is no more but only an extrapolation.

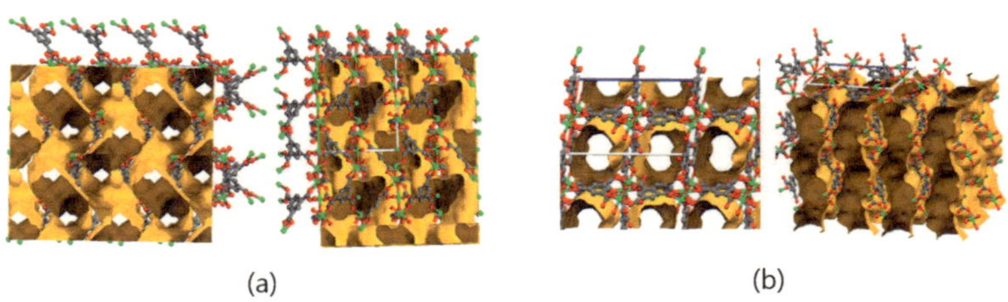

Figure 3.
Diagrams depicting the pores of (a) the A-class topology and (b) the B-class topology at Ni-BTC system. The linear shaped channels of the B-class topology become evident [7].

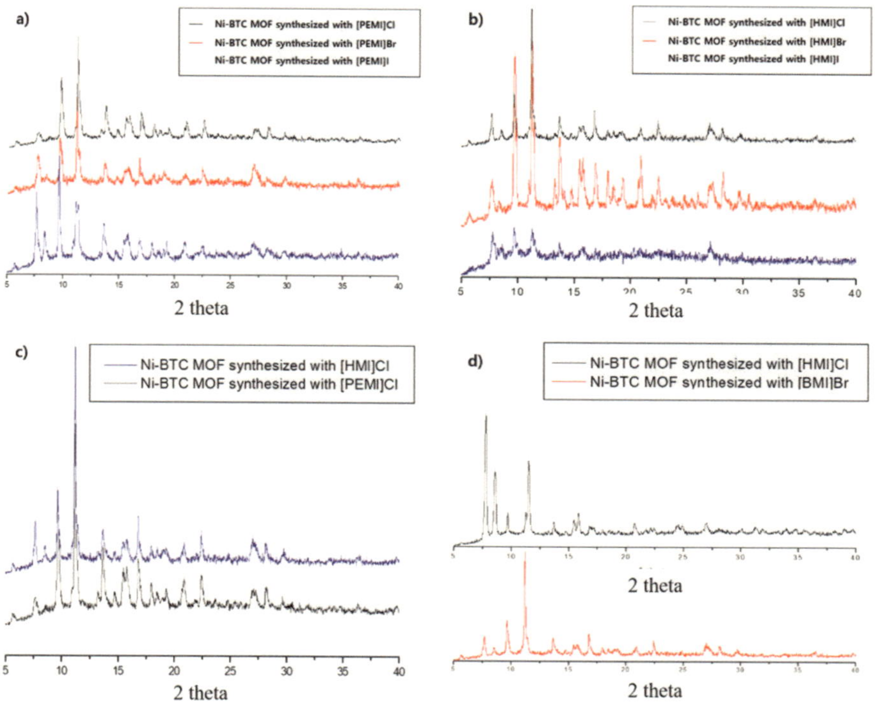

Figure 4.
PXRD data presented in pairs or triplets to illustrate the similarity and difference of the topologies occurring in the Ni-BTC system. (a) Structures synthesised from [PEMI]Cl, [PEMI]Br, and [PEMI]I show that all three in the [PEMI] row belong to the same topology group. (b) Three structures from [HMI] row show that they belong to the same topology group similarly to (a). (c) Two structures from the Cl column show that the topology groups of [HMI] row and [PEMI] row are identical. (d) The structure synthesised from [HMI]Cl is compared to that from [BMI]Br and shows that each two belong to the same class. The magnitude may be different, but the positions of the main peaks coincide [7].

3.2 The effect of the final framework on the solvent properties

The trend extrapolation introduced above was a success, but by no means is a guarantee that similar arguments will always hold true. All scientific explanations are based on reductive models where many details in reality must have been missed. However, discussions until now have only focused on how the variance in solvent properties, namely size, give rise to another variance in the product framework, but never vice versa. Chemistry is a study of interactions, and the term 'interactions' imply that there may always be bidirectional influence. It surely is the solvent that was in the reaction system first and then frameworks were built on top of the influence of the solvent, and thus, its influence on the product is more pronounced and also more important. However, the ionic species residing in different structures are in fact different even if they were in the same bottle before being deployed to the reaction. We would now like to guide your attention to an interesting system where the framework exerts a strong pressure to the cation to alter its shape. In such cases, extrapolations may not give accurate results.

Table 6 shows three reported entries in the not-so-extensively-studied cobalt-BTC system. Nevertheless, simplicity is not to be confused with incompleteness. While it may be true that the entire trend may not be fully described, deeper analysis may follow in a more complete manner for the part that has been, or at least it reduces the burden to describe so many entries in full details.

The bromide column is never an exception to the systems we have been through. The α-class topology has pore volume smaller than the β-class, and it is

$Co(OAc)_2 \cdot H_3BTC$	Cl	Br	I
[EMI]	—	α1 [25]	—
[PMI]	—	β1 [30]	—
[BMI]	—	β2 [31]	—

Each labels denote, α1-$[EMI]_2[Co_3(BTC)_2(OAc)_2]$, β1- $[PMI]_2[Co_2(BTC)_2(H_2O)_2]$, and β2 -$[BMI]_2[Co_2(BTC)_2(H_2O)_2]$. Combinations that have not been reported were left blank.

Table 6.
*A table for the system of framework synthesised with cadmium-BTC system arranged similarly to **Table 1**.*

the increase in the cation size that caused it. However, a question that never has been addressed in previous systems was, is the difference between [EMI] to [PMI] the same as that between [PMI] and [BMI]? In other words, they are gradual, but are they in scale? They both differ by a carbon, and carbon–carbon bond length is nearly universally conserved. It seems they should differ only by an iota, if they were even different after all.

The *in situ* conformations of the guest cations were taken and subjected to computational analysis [31]. The difference in volume between [EMI] and [PMI] was calculated to be 21.8 $Å^3$, which is significantly larger than 14.9 $Å^3$, the difference between [PMI] and [BMI]. It is apparent that this difference arose from the bent conformation of the butyl chain of [BMI] cation; the distance between the terminal carbon to the first carbon in the chain was 2.918°A in compound β2, exceeding 2.567°A of compound β1 only by a small difference. The carbon–carbon bond is free to rotate about each other, but the β-class framework is stable enough to fix the conformation severely bent as they appear; a remarkable example of the framework influencing the property of the solvent. Moreover, just because it appears as the same one step on the table does not mean the actual size difference between the ionic species is the same.

Even though β1 and β2 structures belong to the same topology class, they may have minor differences like the ones described in the manganese system. Even by a small bit, [BMI] is still larger than [PMI] and is expected to exert pressure on the framework towards retaining a larger void volume. However, this straightforward prediction is actually far from the truth. The β-topology framework is so rigid that the void volume and the framework volume stay nearly unchanged for [PMI] and [BMI].

It also deserves some attention that the β-topology occurs very rare in other metal systems, suggesting that it is not so chemically favoured in many other environments [31]. While the rigidity of the framework can also be viewed as how favoured it is over other possible outcomes, it is interesting that this rare topology is so strongly preferred in the system and in cobalt system only. Also, attempts to synthesise crystalline frameworks with [PEMI]Br and [HMI]Br in the system all failed but only acquired amorphous solids. This further supports the absence of any other stable framework possible in the cobalt system. Additional studies must follow to provide explanations for the strikingly different preference of framework in the cobalt system.

4. Reducing topologies can easily deliver deep insights into the structures

The structural details of nanoscopic frameworks are often difficult to perceive. Some basic discussion may be made even with the structures completely ignored, but we already have seen many limitations to that. Understanding the structure is necessary to provide more thorough explanations for the chemical trends appearing in the organised systems of ionothermally prepared MOFs, including many unusual cases unexplainable by simple intuition. Just like organic chemistry cannot be approached without molecular formulas, inorganic chemistry cannot be explained without framework structures. We would like to dedicate this chapter to suggest a method to break down the complications of nanoscopic structures to see the forest beyond the trees, and lastly, tour around that forest.

4.1 Metals atoms tend to exist in clusters

In order to bring down the structures to simpler diagrams, the patterns, or segments of atoms, that occur frequently throughout the framework must be well noticed. After taken the knowledge of the building blocks, we will look into a representative building to see how the blocks are assembled to a building. It is obvious that the organic linker will stay as it is used before the reaction in most structures, as it is very difficult for the benzene ring to disassemble in our BTC example. One thing, however, may fluctuate greatly from structure to structure: the coordination mode. Often there are many atoms, or sites, that are capable of coordinating to metal atoms, but almost always, not all of them do. It is very difficult to predict which coordination mode the ligand will take, since even under the same topology, the ligands are found to take structures with many different coordination modes [1, 30, 31]. Attempts have been made to collectively study coordination modes [34], but for successful discovery of any laws governing them, acquisition of more data is necessary.

In collaboration with the coordination modes, though it is difficult to distinguish causation from correlation at this level, the reaction environment determines the shape in which the metal atoms exist in the framework. From **Tables 4**, 5 and 7, it has been shown the nuclear types the metal atoms take in the framework, but the concept has never been visited yet. This 'nucleus' is a small collection of metal atoms and atoms from the organic ligand coordinating to them and is more commonly called 'metal clusters' because many metal atoms are found together in most structures. These metal clusters are one of the most important character to determine the topology of MOFs, and the frameworks are named as binuclear, trinuclear, etc. according to the number metal atoms present in the metal cluster. If small variations within the same topology are ignored, the framework can be viewed as a collection of simple connections between the unvarying organic ligand and the metal clusters, just like vertices and edges of a mathematical 3D figure.

The simplification illustrated in **Figure 5** exemplifies the power of reduction in brining different structures together. Although it could have been inferred from the same molecular formulas, a great number of structures introduced in **Tables 4**, **5** and 7 actually have the exact same framework.

4.2 Structure explains the popularity of [RMI][metal(BTC)] topology

Some of the most commonly occurring structures need attention, not only because they will be frequently met in trials of novel conditions, but also they will provide a valuable starting point in relating to other structures occurring in the same system to understand the correlations like the ones we have visited.

The topology [RMI][Metal(BTC)] occurs in most metal systems that have been reported and in the highest frequency. With this topology as an example, we will show how a complex structure may be simplified. This way, details unnecessary for understanding of the topology can be ignored and attention may be more easily focused on the topology itself. The characteristics that may vary within the topology without changing it include coordination modes, bond angle, and bond in certain ranges.

M	Formula	Nuclear #	CCDC code	Void with cation*	Void without cation*
Co	[EMI][Co(1,4-ndc)Br]	Di-nuclear	AHIYIB [37]	0% 0 Å^3 /1739.88 Å^3	37.8% 658.13 Å^3 /1739.88 Å^3
	$[PMI]_2[Co_7(1,4\text{-}ndc)_6(OH)_4]$	Hepta-nuclear	AHIYOH [37]	0% 0 Å^3 /8376.27 Å^3	23.4% 1958.42 Å^3 /8376.27 Å^3
	$[BMI]_2[Co_6(1,4\text{-}ndc)_6(OH)_2]$	Hexa-nuclear	AHIYUN [37]	0% 0 Å^3 /7998.16 Å^3	21.5% 1717.36 Å^3 /7998.16 Å^3
	$[AMI]_4[Co_4Na_5(1,4\text{-}ndc)_8Br]$	Tetra-nuclear	AHIZAU [37]	0% 0 Å^3 /6065.78 Å^3	31.3% 1897.74 Å^3 /6065.78 Å^3
Eu	$[EMI][Eu(1,4\text{-}ndc)(ox)_{0.5}Br]$	Poly-nuclear	EMUTUD [38]	0% 0 Å^3 /2033.18 Å^3	34.8% 708.05 Å^3 /2033.18 Å^3
	$[HMI][Eu_2Cl(1,4\text{-}ndc)_3]$	Poly-nuclear	MEGPIZ [39]	0% 0 Å^3 /4188.58 Å^3	26.0% 1086.94 Å^3 /4188.58 Å^3
Sm	$[HMI][Sm_2Cl(1,4\text{-}ndc)_3]$	Poly-nuclear	MEGPEV [39]	0% 0 Å^3 /4203.36 Å^3	25.8% 1086.30 Å^3 /4203.36 Å^3
	$[EMI][Sm(1,4\text{-}ndc)(ox)_{0.5}Br]$	Poly-nuclear	EMUVAL [38]	0% 0 Å^3 /2032.62 Å^3	34.9% 709.04 Å^3 /2032.62 Å^3
Mg	$[AMI]_2[Mg_3(1,4\text{-}ndc)_4(MeIm)_2(H_2O)_2]\cdot H_2O$	Tri-nuclear	EXUYEC [40]	0% 0 Å^3 /3424.65 Å^3	24.7% 845.14 Å^3 /3424.65 Å^3
	$[BMI]_2[Mg_6(1,4\text{-}ndc)_5(H_2NDC)_2(HCOO)_2]$	Hexa-nuclear	NUKBOM [41]**	0% 0 Å^3 /4700.01 Å^3	0% 0 Å^3 /4700.01 Å^3
Zn	$[EMI]_2[Zn_7(\mu_4\text{-}O)_2(1,4\text{-}ndc)_6]$	Hepta-nuclear	QATLEE [42]	0% 0 Å^3 /8044.7 Å^3	23.2% 1864.05 Å^3 /8044.7 Å^3
Ni	$[Ni_3(1,4\text{-}ndc)_4(Mim\text{-}C3Im)_2(H_2O)_2]$	Di-nuclear	TOLMOY [43]	0% 0 Å^3 /1704.13 Å^3	36.5% 622.27 Å^3 /1704.13 Å^3
	$[Ni_3(1,4\text{-}ndc)_4(Mim\text{-}C4Im)_2(H_2O)_2]$	Di-nuclear	TOLMUE [43]	0% 0 Å^3 /1721.79 Å^3	37.3% 642.55 Å^3 /1721.79 Å^3
	$[Ni_3(1,4\text{-}ndc)_4(Mim\text{-}C5Im)_2(H_2O)_2]$	Tri-nuclear	TOLNAL [43]**	0% 0 Å^3 /3276.26 Å^3	0% 0 Å^3 /3276.26 Å^3
	$[Ni_3(1,4\text{-}ndc)_4(Mim\text{-}C6Im)_2(H_2O)_2]$	Tri-nuclear	TOLNEP [43]**	0% 0 Å^3 /3387.26 Å^3	0% 0 Å^3 /3387.26 Å^3
Cd	[EMI][CdBr(1,4-ndc)]	Di-nuclear	UYUPUA [44]	0% 0 Å^3 /1814.56 Å^3	38.4% 697.55 Å^3 /1814.56 Å^3
Sr	$[EMI][Sr_{10}(1,4\text{-}ndc)_{10}Br_4]$	Poly-nuclear	VUXGOM [45]	0.2% 17.02 Å^3 /6950.81 Å^3	18.4% 1279.20 Å^3 /6950.81 Å^3

M	Formula	Nuclear #	CCDC code	Void with cation*	Void without cation*
Na, Cu	[EMI][NaCu(1,4-ndc)$_2$]	Poly-nuclear	WIKZAT [44]	0.6% 16.45 Å^3 /2729.18 Å^3	25.7% 700.27 Å^3 /2729.18 Å^3
La	[HMI][La$_2$Cl(1,4-ndc)$_3$]	Poly-nuclear	MEGNIX [39]	0% 0 Å^3 /4328.43 Å^3	26.6% 1150.84 Å^3 /4328.43 Å^3
Ce	[HMI][Ce$_2$Cl(1,4-ndc)$_3$]	Poly-nuclear	MEGNOD [39]	0% 0 Å^3 /4296.06 Å^3	26.6% 1142.24 Å^3 /4296.06 Å^3
Pr	[HMI][Pr$_2$Cl(1,4-ndc)$_3$]	Poly-nuclear	MEGNUJ [39]	0% 0 Å^3 /4262.42 Å^3	26.3% 1122.11 Å^3 /4262.42 Å^3
Nd	[HMI][Nd$_2$Cl(1,4-ndc)$_3$]	Poly-nuclear	MEGPAR [39]	0% 0 Å^3 /4241.32 Å^3	26.4% 1121.26 Å^3 /4241.32 Å^3
Gd	[HMI][Gd$_2$Cl(1,4-ndc)$_3$]	Poly-nuclear	MEGPOF [39]	0% 0 Å^3 /4164.69 Å^3	25.7% 1069.98 Å^3 /4164.69 Å^3
Tb	[HMI][Tb$_2$Cl(1,4-ndc)$_3$]	Poly-nuclear	MEGPUL [39]	0% 0 Å^3 /4152.78 Å^3	25.7% 1066.29 Å^3 /4152.78 Å^3
Dy	[HMI][Dy$_2$Cl(1,4-ndc)$_3$]	Poly-nuclear	MEGQAS [39]	0% 0 Å^3 /4124.91 Å^3	25.5% 1052.86 Å^3 /4124.91 Å^3
Ho	[HMI][Ho$_2$Cl(1,4-ndc)$_3$]	Poly-nuclear	MEGQEW [39]	0% 0 Å^3 /4105.46 Å^3	25.4% 1044.15 Å^3 /4105.46 Å^3
Er	[HMI][Er$_2$Cl(1,4-ndc)$_3$]	Poly-nuclear	MEGQIA [39]	0% 0 Å^3 /4103.22 Å^3	25.4% 1040.46 Å^3 /4103.22 Å^3
Y	[HMI][Y$_2$Cl(1,4-ndc)$_3$]	Poly-nuclear	MEGQOG [39]	0% 0 Å^3 /4127.12 Å^3	25.5% 1052.17 Å^3 /4127.12 Å^3

Calculation was done with the same setting. In the entry with reference code NUKBOM, TOLNAL, and TOLNEP, the calculation for the void volume with the cation removed was not conducted since the cations were bound to the framework.

Table 7.
*Structures arising from imidazolium-based MOFs with 1,4-naphthalene dicarboxylic acid(NDC) presented in a similar manner to **Table 4**.*

The simplification above is itself beautiful but is meaningless if description of the topology is not accompanied. Description gives meaning to the structure and explanations for many of the observed phenomena.

Based on face-centred cubic lattice (FCC), the unit cell of [RMI][Metal(BTC)] is very compact. Its binuclear metal cluster occupies all the FCC sites, and BTC occupies the interstitial sites. There are eight BTC ligands, and the rest of the interstitial sites appear empty in **Figure 6**. These sites, however, are not actually empty. There are eight metal clusters and eight BTC ligands in the unit cell, but each metal cluster has double positive charge while BTC ligand has triple negative. The framework is negatively charged, as nearly every ionothermally synthesised framework is, and the charge balance is maintained by the guest cations occupying the rest of the

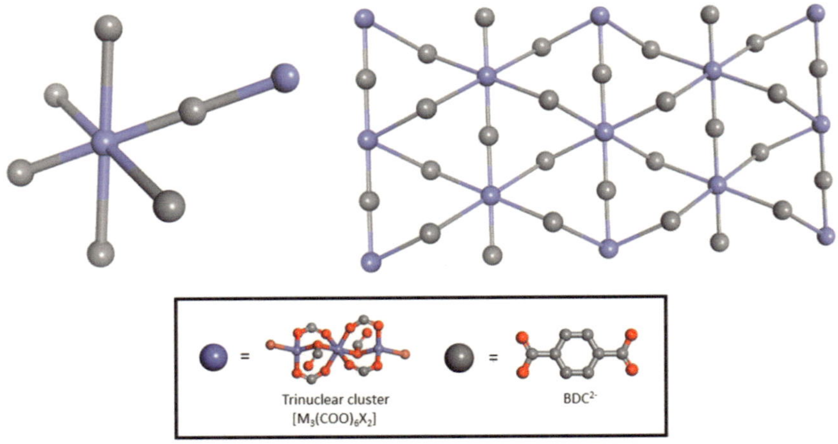

Figure 5.
The structure represents the (2,6)-connected 2D network. The list of entries that exhibit this particular structure is: $[EMI]_2[Co_3(BDC)_3X_2]$ (X = Cl, Br, I) (TACHUD for Cl; JOQXOE for Br; TACJOZ for I), $[PMI]_2[Co_3(BDC)_3X_2]$ (X = Cl, Br, I) (TACJAL for Cl; JOQXUK for Br; TACJUF for I), $[BMI]_2[Co_3(BDC)_3X_2]$ (X = Cl, Br, I) (TACJEP for Cl; JOQPUC for Br; TACKAM for I), $[PEMI]_2[Co_3(BDC)_3X_2]$ (X = Cl, I) (TACJIT for Cl; TACKEQ for I), $[EMI]_2[Zn_3(BDC)_3X_2]$ (X = Cl, Br, I) (SIVQEV for Cl; QUGVIZ for Br; QUGWEW for I), $[PMI]_2[Zn_3(BDC)_3X_2]$ (X = Cl, Br, I) (QUGVAR for Cl; QUGVOF for Br; QUGWIA for I), $[BMI]_2[Zn_3(BDC)_3X_2]$ (X = Cl, Br, I) (SIVCAD for Cl; QUGVUL for Br; QUGWOG for I), and $[PEMI]_2[Zn_3(BDC)_3X_2]$ (X = I) (QUGWUM for I).

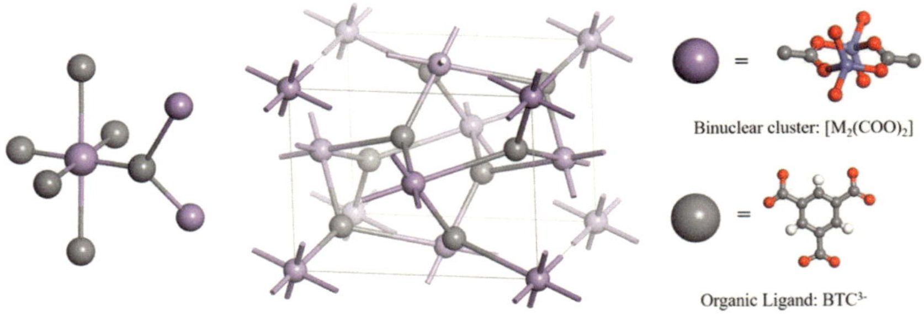

Figure 6.
The asymmetrical unit and the unit cell of [RMI][Metal(BTC)] topology with its binuclear cluster and organic ligand represented, respectively, by singular units.

interstitial sites. This allows no void for the structure and is thus stable. Nevertheless, the structure may not house longer cations regardless of how preferred it is over other possible options; it is just impossible. This complies with the observation from **Table 5** that the structure is very much preferred with [EMI], but only with [EMI] and the preference drops greatly as we move on to longer cations.

4.3 Seemingly different structures may belong to the same topology class

A large number of syntheses have been reported to the literature, but the number of novel topologies is much smaller. It will be very interesting to see so many structures that once appeared all different converging into one topology. In this example, a group of structures with a different formula and a different nuclear type will be merged with the [RMI][Metal(BTC)] class that has been described above.

Figure 7 depicts the [RMI][Metal(BTC)] structure. This same structure, however, is shared by $[EMI]_2[In_2Co(OH)_2(BTC)_2Br_2]$ structure that has a remarkably different molecular formula. The formula is the simplest tool to represent frameworks, but it can sometimes be misleading. **Figure 8** shows an even more striking

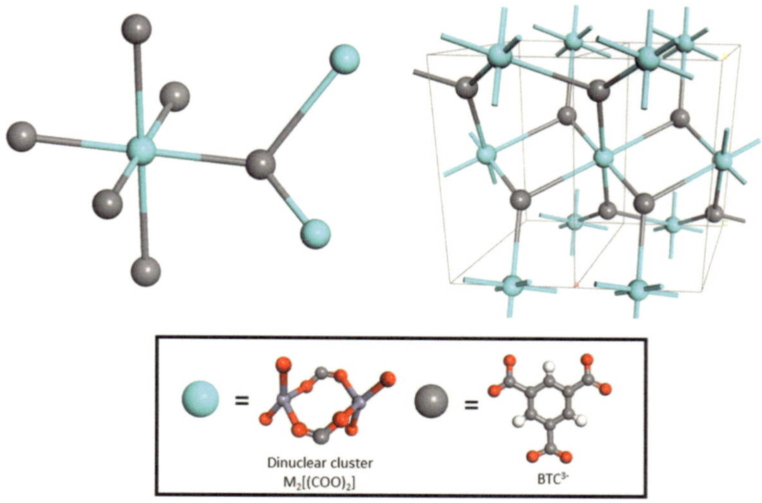

Figure 7.
The structure represents (3,6)-connected network. This corresponds to the following formulas: [BMI][Zn(BTC)] (FUTZEB), and $[EMI]_2[In_2Co(OH)_2(BTC)_2Br_2]$ (VUVZAP).

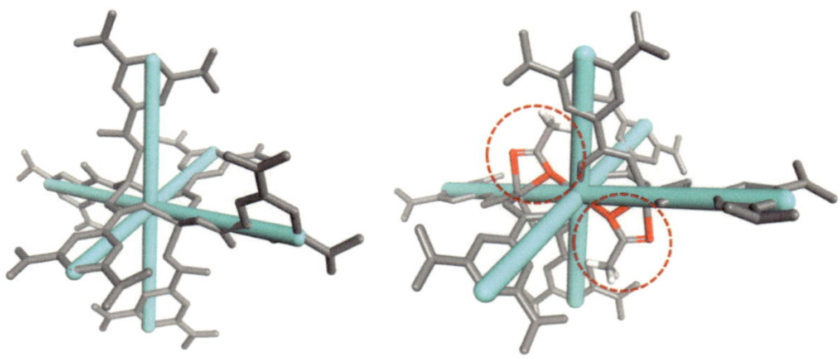

Figure 8.
The left is the [RMI][Metal(BTC)] class with binuclear clusters and the right $[RMI]_2[Metal_3(HBTC)_4(H_2O)]$ with trinuclear clusters. Light blue stick represents the octahedral connections between metal clusters and BTC to better illustrate the topological identity of the two groups. The molecular moieties in the red circle represent acetate groups.

example with [RMI][Metal(BTC)] structure and $[RMI]_2[M_3(HBTC)_4(H_2O)]$ structure. Although it is very difficult to catch any similarities from the formula nor the structure if at first glance they actually fall under the same topology umbrella. This remarkable similarity is possible because some coordination sites of the trinuclear metal cluster are occupied by another molecular moiety, OAc in this case as shown in **Figure 7**. These places the trinuclear clusters in the octahedral coordination mode, which is the maximum coordination that binuclear metal clusters can have.

5. Outstanding properties of ionothermally prepared MOFs

In previous sections, we have explored through the diverse structures prepared by ionothermal synthesis and several perspectives through which the groups of structures may be analysed to gain deeper insights. The last step is to find a practical use for those insights. The versatility of ionothermal synthesis, that its reaction environment may be easily altered and related to the change in products, directly leads us towards the diversity of structures that may be prepared through the methodology. As such, ionothermal synthesis promises a variety of potential uses, although most of them have obstacles yet to be resolved along the way towards practical employment.

5.1 Ion exchange is the key to make use of the pores

Many of the most popularly studied application of MOFs make use of the frameworks as molecular sieves. The nanoscale pores of MOFs can selectively filter out any chemicals that do not fit into them and this selectivity can be chosen by the industry among the diversity of reported structures. The first use of ionothermally prepared MOFs is probably also the same. In this case, ionothermal synthesis has one advantage that the solvent functions as a template and can be varied in size to modify the pore size. However, it is a double-edged sword that actually limits the practicality of ionothermal synthesis. To make any use of the pores, the templates occupying the pores must be removed. The problem is that they hardly ever do.

The void volume of the structures synthesised with the cation varied in size has been compared in **Tables 4, 5** and 7. Frameworks with the solvent resident in their channels, or cavities, tend to have compact structure with the void volume as low as 0% of the unit cell volume. For your reference, MOF-5, a representative framework, has a void cavity as large as 70% of unit cell volume. This absence of void volume arose because of the large solvent cations stuck in the cavity, rather than the framework itself. When calculated with the resident cations completely removed, void volume was increased to approximately 50% of unit cell volume. In theory, the large volume occupied by the cations may be decreased by subjecting the framework to ion exchange with smaller cations, so that the rest may be used purposefully. Unfortunately for now, this possibility seems to stay only in theory. Given its important position—the first step in bringing ionothermal synthesis to practicality, tremendous efforts have been put into making this exchange possible, but they rarely succeeded. In one case that we tested, evacuation of cations was observed in $[BMI]_2[Co_2(BTC)_2(H_2O)_2]$ crystals upon treatment with water, but only when accompanied with significant destruction of the framework [31]. Nevertheless, Li et al. reported partial but stable ion exchange with $[EMI]_2[In_2Co(OH)_2(BTC)_2Br_2]$ crystals [32], suggesting a new possibility for the ionothermal synthesis methodology.

5.2 Placing metal atoms in proximity to yield novel characteristics

The limitations posed by the irreplaceable templates have indeed disappointed the researchers and presumably many of you, too. However, even if the pores of ionothermally prepared MOFs are totally unusable, they still have some valuable characteristics. It is very common in the world of nanoscience that a substance acquires some characteristics completely different from those of its macroscopic bulk. One of the most frequently reported application is detection of chemicals via photoluminescence that changes upon encounter with specific chemicals. This includes the photoluminescence of europium ions in $[HMI][Eu(DHBDC)_2$, where DHBDC indicates 2,5-dihydroxytelephtalic acid, capable of detect Ba^{2+} ions quantitatively [25], and $[RMI][Eu_2(BDC)_3Cl]$ for detection of aniline [18]. In addition, ionothermally prepared $[EMI][Dy_3(BDC)_5]$, a rod-shaped polymer, has been shown to exhibit slow magnetic relaxation behaviour like single-molecule magnets [22]. It seems like ionothermally prepared structures may be applied for any purposes that exist in the field of nanochemistry.

6. Conclusion

In a system of different chemical outcomes from related starting conditions, it is often difficult to track what has caused that difference. The ionothermal synthesis methodology is excellent in this aspect. Changes can be made in a gradual and continuous manner to observe how the result reacts to them. By organising the results based on their solvents, we can connect the dots to find useful correlations that can be both used for intra- and extrapolation. The relations are often very intuitive. For this simplicity to not lead to inaccuracy, there is a need to carefully examine the frameworks and how they react with the reaction environment—the ionic liquid that functions both as solvent and template. In the course of simplifying the frameworks for examination follows merging of structures into large topology groups, which can be used to better organise once separated correlations from various chemical systems. Despite of all the positive characters of ionothermal synthesis, however, there is a limit to their practical application—the irreplaceable reaction templates. For this rich and potentially richer field of material synthesis to contribute to the industry, efforts must be made to either resolve the issue or to find uses from the framework itself, rather than the voids.

List of abbreviations

Cations	
RMI	1-alkyl-3-methylimidazolium
EMI	1-ethyl-3-methylimidazolium
PMI	1-propyl-3-methylimidazolium
BMI	1-butyl-3-methylimidazolium
PEMI	1-pentyl-3-methylimidazolium
HMI	1-hexyl-3-methylimidazolium
Organic ligands	
BTC	1,3,5-benzenetricarboxylic acid
BDC	1,4-benzene dicarboxylic acid
NDC	1,4-naphthalene dicarboxylic acid
4,4′-bpy	4,4′-bipyridine
DHBDC	2,5-dihydroxytelephtalic acid
Others	
MOF	metal organic framework
CSD	Cambridge structural database
PXRD	powder X-ray diffraction
OAc	acetate
CCDC	Cambridge crystallographic data centre

Author details

Hyun-Chang Oh[1], Sukwoo Jung[2], Il-Ju Ko[1] and Eun-Young Choi[1,2]*

1 Department of Chemistry, KAIST, Daejeon, Republic of Korea

2 Department of Chemistry and Biology, Korea Science Academy of KAIST, Busan, Republic of Korea

*Address all correspondence to: faujasite1@kaist.ac.kr

References

[1] Xu L, Yan S, Choi E-Y, Lee JY, Kwon Y-U. Product control by halide ions of ionic liquids in the ionothermal syntheses of Ni-(H)BTC metal-organic frameworks. Chemical Communications. 2009:3431-3433

[2] Jin K et al. [Cu(i)(bpp)]BF4: The first extended coordination network prepared solvothermally in an ionic liquid solvent. Chemical Communications. 2002:2872-2873

[3] Morris RE, Weigel SJ. The synthesis of molecular sieves from non-aqueous solvents. Chemical Society Reviews. 1997;**26**:309-317

[4] Del Pópolo MG, Voth GA. On the structure and dynamics of ionic liquids. The Journal of Physical Chemistry. B. 2004;**108**:1744-1752

[5] Rogers RD, Seddon KR. Ionic liquids—solvents of the future? Science. 2003;**302**:792-793

[6] Earle MJ et al. The distillation and volatility of ionic liquids. Nature. 2006;**439**:831-834

[7] Hyun Chang O, Sukwoo J, Ko I-J, Eun Young C. Tabular organisation of ionothermally prepared MOFs to extrapolate chemical trends and successfully predict synthesis results. Biomedical Journal of Scientific & Technical Research. 2018). BJSTR. MS.ID.000996;**4**(1):4. DOI: 10.26717/BJSTR.2018.04.000996

[8] Plechkova NV, Seddon KR. Ionic liquids: "Designer" solvent for green chemistry. In: Tundo P, Perosa A, Zecchini F, editors. Methods and Reagents for Green Chemistry. Wiley. 2007. DOI: 10.1002/9780470124086.ch5

[9] Tokuda H, Hayamizu K, Ishii K, Susan MABH, Watanabe M. Physicochemical properties and structures of room temperature ionic liquids. 2. Variation of alkyl chain length in imidazolium cation. The Journal of Physical Chemistry. B. 2005;**109**:6103-6110

[10] Xu L, Kwon Y-U, de Castro B, Cunha-Silva L. Novel Mn(II)-based metal-Organic frameworks isolated in ionic liquids. Crystal Growth & Design. 2013;**13**:1260-1266

[11] Parnham ER, Morris RE. Ionothermal synthesis of zeolites, metal-organic frameworks, and inorganic-organic hybrids. Accounts of Chemical Research. 2007;**40**:1005-1013

[12] Xu L, Choi EY, Kwon YU. Combination effects of cation and anion of ionic liquids on the cadmium metal-organic frameworks in ionothermal systems. Inorganic Chemistry. 2008;**47**:1907-1909

[13] Liao J-H, Wu P-C, Huang W-C. Ionic liquid as solvent for the synthesis and crystallization of a coordination polymer: (EMI)[Cd(BTC)] (EMI = 1-Ethyl-3-methylimidazolium, BTC = 1,3,5-benzenetricarboxylate). Crystal Growth & Design. 2006;**6**:1062-1063

[14] Zhang Z-H, Xu L, Jiao H. Ionothermal synthesis, structures, properties of cobalt-1,4-benzenedicarboxylate metal–organic frameworks. Journal of Solid State Chemistry. 2016;**238**:217-222

[15] Xu L et al. The influence of 1-alkyl-3-methyl imidazolium ionic liquids on a series of cobalt-1,4-benzenedicarboxylate metal–organic frameworks. CrystEngComm. 2014;**16**:10649-10657

[16] Tapala W, Prior TJ, Rujiwatra A. Two-dimensional anionic zinc

benzenedicarboxylates: Ionothermal syntheses, structures, properties and structural transformation. Polyhedron. 2014;**68**:241-248

[17] Zhang Z-H, Liu B, Xu L, Jiao H. Combination effect of ionic liquid components on the structure and properties in 1,4-benzenedicarboxylate based zinc metal–organic frameworks. Dalton Transactions. 2015;**44**:17980-17989

[18] Feng H-J, Xu L, Liu B, Jiao H. Europium metal-organic frameworks as recyclable and selective turn-off fluorescent sensors for aniline detection. Dalton Transactions. 2016;**45**:17392-17400

[19] Cao HY et al. Ionothermal syntheses, crystal structures and luminescence of three three-dimensional lanthanide-1,4-benzenedicarboxylate frameworks. Inorganica Chimica Acta. 2014;**414**:226-233

[20] Huang G, Yang P, Wang N, Wu JZ, Yu Y. First lanthanide coordination polymers with N,N-dimethylformamide hydrolysis induced formate ligands. Inorganica Chimica Acta. 2012;**384**:333-339

[21] Li S-Y, Du L, Liu Z-H. Ionothermal synthesis, crystal structure, and luminescent properties of two novel layered indium-1,4-benzenedicarboxylate complexes. Synthesis and Reactivity in Inorganic, Metal-Organic, and Nano-Metal Chemistry. 2016;**46**:675-680

[22] Liu Q-Y et al. Ionothermal synthesis of a 3D dysprosium-1,4-benzenedicarboxylate framework based on the 1D rod-shaped dysprosium-carboxylate building blocks exhibiting slow magnetization relaxation. CrystEngComm. 2014;**16**:486-491

[23] Liao J-H, Huang W-C. Ionic liquid as reaction medium for the synthesis and crystallization of a metal-organic framework: $(BMIM)_2[Cd_3(BDC)_3Br_2]$ (BMIM=1-butyl-3-methylimidazolium, BDC=1,4-benzenedicarboxylate). Inorganic Chemistry Communications. 2006;**9**:1227-1231

[24] Xiahou Z-J, Wang Y-L, Liu Q-Y, Wei J-J, Chen L-L. Ionothermal synthesis of a 3D heterometallic coordination polymer based on the rod shaped copper(II)–sodium(I)-carboxylate secondary building units with a pcu topology. Inorganic Chemistry Communications. 2013;**38**:62-64

[25] Lin Z, Wragg DS, Morris RE. Microwave-assisted synthesis of anionic metal–organic frameworks under ionothermal conditions. Chemical Communications. 2006:2021-2023. DOI: 10.1039/B600814C

[26] Lin Z, Slawin AMZ, Morris RE. Chiral induction in the ionothermal synthesis of a 3-D coordination polymer. Journal of the American Chemical Society. 2007;**129**:4880-4881

[27] Xu L, Choi E-Y, Kwon Y-U. A new 3D nickel(II) framework composed of large rings: Ionothermal synthesis and crystal structure. Journal of Solid State Chemistry. 2008;**181**:3185-3188

[28] Lin Z, Li Y, Slawin AMZ, Morris RE. Hydrogen-bond-directing effect in the ionothermal synthesis of metal coordination polymers. Dalton Transactions. 2008:3989-3994. DOI: 10.1039/b802892c

[29] Lin Z, Wragg DS, Warren JE, Morris RE. Anion control in the ionothermal synthesis of coordination polymers. Journal of the American Chemical Society. 2007;**129**:10334-10335

[30] Wang YL et al. Ionothermal syntheses and crystal structures of

two cobalt(II)-carboxylate compounds with different topology. Inorganic Chemistry Communications. 2011;**14**:380-383

[31] Ko IJ, Oh HC, Cha YJ, Han CH, Choi EY. Ionothermal synthesis of a novel 3D cobalt coordination polymer with a uniquely reported framework: $[BMI]_2[Co_2(BTC)_2(H_2O)_2]$. Advances in Materials Science and Engineering. 2017;**2017**:6

[32] Li SY, Liu ZH. Synthesis, structure and property of a 3D heterometallic complex constructed by trinuclear $[In_2Co(OH)_2(COO)_4]$ cluster and BTC ligand. Journal of Cluster Science. 2015;**26**:1959-1970

[33] Xu L, Choi EY, Kwon YU. Ionothermal synthesis of 3D zinc coordination polymer: $[Zn_2(BTC)(OH)(I)](BMIM)$ containing novel tetra nuclear building unit. Inorganic Chemistry Communications. 2008;**11**:150-154

[34] Xu L, Choi EY, Kwon YU. Ionothermal syntheses of six three-dimensional zinc metal-organic frameworks with 1-alkyl-3-methylimidazolium bromide ionic liquids as solvents. Inorganic Chemistry. 2007;**46**:10670-10680

[35] Xu L, Choi EY, Kwon YU. Ionothermal synthesis of a 3D Zn-BTC metal-organic framework with distorted tetranuclear $[Zn_4(\mu_4\text{-}O)]$ subunits. Inorganic Chemistry Communications. 2008;**11**:1190-1193

[36] Ordonez C, Fonari M, Lindline J, Wei Q, Timofeeva T. How structure-directing cations tune the fluorescence of metal-organic frameworks. Crystal Growth & Design. 2014;**14**:5452-5465

[37] An B, Wang J-L, Bai Y, Dang D-B. Systematic design of secondary building units by an efficient cation-directing strategy under regular vibrations of ionic liquids. Dalton Transactions. 2015;**44**:14666-14672

[38] Gao MJ, Wang YL, Cao HY, Liu QY, Chen LL. Ionothermal syntheses, crystal structures and luminescence of two lanthanide-carboxylate frameworks based on the 1, 4-naphthalenedicarboxylate and oxalate mixed ligands. Zeitschrift für anorganische und allgemeine Chemie. 2014;**640**:2472-2476

[39] Tan B et al. Ionothermal syntheses, crystal structures and properties of three-dimensional rare earth metal-organic frameworks with 1,4-naphthalenedicarboxylic acid. Dalton Transactions. 2012;**41**:10576-10584

[40] Wu ZF, Hu B, Feng ML, Huang XY, Zhao YB. Ionothermal synthesis and crystal structure of a magnesium metal-organic framework. Inorganic Chemistry Communications. 2011;**14**:1132-1135

[41] Wu Z-F et al. An ionothermally synthesized Mg-based coordination polymer as a precursor for preparing porous carbons. CrystEngComm. 2015;**17**:4288-4292

[42] Wei JJ, Liu QY, Wang YL, Zhang N, Wang WF. Ionothermal synthesis of a 3D zinc(II)-carboxylate coordination polymer with bcu topology based on heptanuclear $[Zn_7(\mu_4\text{-}O)_2]$ cluster. Inorganic Chemistry Communications. 2012;**15**:61-64

[43] An B, Bai Y, Wang J-L, Dang D-B. Cation-size-controlled assembly of the Ni(Ac)2-1, 4-H2NDC system: Geminal dicationic ionothermal syntheses, crystal structures and magnetic properties. Dalton Transactions. 2014;**43**:12828-12831

[44] Tan B, Xie ZL, Huang XY, Xiao XR. Ionothermal synthesis, crystal

structure, and properties of an anionic two-dimensional cadmium metal organic framework based on paddle wheel-like cluster. Inorganic Chemistry Communications. 2011;**14**:1001-1003

[45] Liu SS, Cheng M, Li B. Ionothermal Synthesis of a 3D Luminescent Strontium(Π) Coordination Polymer with Dodecanuclear Metallocylclic Ring Segments. Journal of Inorganic and Organometallic Polymers. 2015;**25**:1103-1110

Chapter 7

Characterization and Treatment of Real Wastewater from an Electroplating Company by Raw Chitin

Boukhlifi Fatima

Abstract

The objective of this chapter is to study of the heavy metal removal in real waste water. The use of the raw chitin shows itself of big potential for the treatment of the liquid discharges of the studied unity. It showed itself capable to treating heavy metal loads superior to 200 mg/l by presenting percentage removal between 90 and 97%, as in the case of Cu^{2+}. After the study performed on the global discharge, we were interested in the local treatment that rinses out plating baths, and this is the aim to optimize the treatment process and develop a project of treatment plant, recycling in situ based on the adsorption technique on raw chitin. Examination of the results allowed us to save significant percentages of sewage treated for metals mainly copper. Raw chitin showed a high affinity toward heavy metals in rinsing water supply. According to this study, the design of a treatment facility of this type of release must include a waste water treatment by adsorption on chitin. The valuation of the raw chitin is situated in this context as an economically adsorbing material, which can be an interest at the level of the recovery of heavy metals in waste water.

Keywords: chitin, adsorption, heavy metals, current rinsing, global rejection, treatment, electroplating

1. Introduction

Generally, a chain of surface treatment consists of sequences of functions (pretreatment, treatment, finishing), consisted of several posts (treatment bath, rinsing, passivation ...). The composition of industrial waste water of treatment surface contains a lot of heavy metals such Cu, Zn, Ni, Cr, Cd and the other toxic materials such as the alkalines, the acids and the cyanides. More and more it contains several complexing agents which influence the used process Na^+, OH^-, SiO_3 and Na_2CO_3, These products are generated in the operations of rinsing of materials before treatment. A variety of acids, HCl, H_2SO_4, HNO_3, H_3PO_4, $HCrO_3$, etc. is used to eliminate the oxidized films. The rejection of the stage of electroplating deposit contains widely metals and the following anions: BO^{3-}, $CO^{2-}{}_3$, $Cr_2O^{2-}{}_7$, F^-, $PO_4{}^{3-}$, Cl^-, NO^{-3} and $SO_4{}^{2-}$. Generally, wastewater contains many heavy metals. The most important of cations are Cr^{6+}, it is reduced to Cr^{3+}, while CN-is oxidized in the preliminary treatment.

Different treatment techniques for wastewater laden with heavy metals have been developed in recent years both to decrease the amount of wastewater produced and to improve the quality of the treated effluent. Although various treatments such as chemical precipitation, coagulation-flocculation, flotation, ion exchange and membrane filtration can be employed to remove heavy metals from contaminated wastewater, they have their inherent advantages and limitations in application [1].

Chemical precipitation is widely used for the treatment of electroplating wastewater [2, 3]. It consists of adding a base followed by sedimentation. The pH is adjusted to the minimum solubility, so it is difficult to treat multiple metals simultaneously. Coagulation-flocculation has also been employed for heavy metal removal from inorganic effluent [4]. Sorptive flotation has attracted interest in Greece and the USA [5–7] for the removal of non-surface active metal ions from contaminated wastewater. In recent years, ion exchange has also received considerable interest in Italy and Spain [8, 9] as one of the most promising methods to treat heavy metals. Starch xanthate (XA) synthetic polymers resins grafted cellulose natural zeolites are used. This process is particularly effective for the recovery of metals, but the cost of the process is very expensive and often justified only metals are recovered [10]. Due to its convenient operation, membrane separation has been increasingly used recently for the treatment of inorganic effluent. There are different types of membrane filtration such as ultrafiltration, nanofiltration and reverse osmosis. Membrane filtration has used in Taiwan and in South Korea [11, 12].

The adsorption treatment is widely used with activated carbons, this process has several advantages: it is very effective in removing heavy metals even at low pH [13] with different adsorbent materials such as polymers [14–16] and clays [17–19]. Electrolysis is generally used to treat water with high metal content. One of the major obstacles encountered in this technique is the complexity of the environments to be treated that leads to a series of redox reactions [20]. In recent year, the clean Technology constituted preventive actions to review and question the production concept, these actions converge to a common point: Targeting the pollution source rather than its reverse vector example osmosis [21] and zero discharge principle [22, 23].

In general, control of water flow requires the establishment of means for counting and control flow at the entrance of each channel. In addition, it reduces the consumption of reagents and production of sludge treatment plant, which is a significant gain in operating costs. Several techniques exist to recover the flow pollutants at the source to mention a few key principles:

- The decrease in volumes trained by parts by adjusting the drainage time.

- The establishment of additional rinse tanks such as rinsing tank death after degreasing, pickling, hot metal deposition and passivation, it traps the metal salts which are then reassembled in the treatment bath to compensate for evaporation. Rinsing; in which the parts are dipped before and after the cold metal plating baths, it can recover from 30 to 50% of the entrained flow.

- Finally, the development of production lines to reduce pollution is an opportunity to optimize production (questioning of manufacturing ranges, the products used and streamlining the flow of parts) and improves quality and working condition.

The objective of this research was to study the ability of raw chitin to purify wastewater loaded with a mixture of heavy metals. This chapter first focused on the physicochemical characterization and the determination of heavy metals levels in wastewater followed by adsorption treatment. We started the study by the global rejection of the society then by the rinsing waters running at the exit of the electroplating baths. Finally, we have studied the comparison between the efficiency of the treatment of heavy metals by chitin shrimps - Ccre or chitin crab -Ccra thus analyzing the possibilities of in situ treatment and recycling.

2. Materials and methods

A study unit is MAFER located at CASABLANCA in MOROCCO. Its activity is surface treatment. The studied unit of surface treatment consists of five chains. Every chain is determined by the succession of tanks. The average capacity of baths varies between 950 and 1710 l. The majority of baths has a volume of 1440 l and is fed by well water except the bath of metallic deposit and the baths of rundown which are filled by the drinking water. Waters of the baths of the dead rinsing are recycled in the bath of metallic deposit. The water supply of well often matches 1 h a day. Whereas the drinkable water supply is made after draining of bath.

2.1 Preparation of the adsorbent

the shells of Shrimp (Ccre) and the shells of crabs (Ccra) are isolated at first by their mild part (protein) washed in bidistilled water then dried at 100°C during 48 h [17], then crushed and sieved. The size grading is understood between 100 and 125 μm.

2.2 Adsorption test

A mass of the adsorbent is placed in contact with the rejection to be treated (100 ml). The suspensions were stirred (500 rev/min) in constant temperature (25 ± 2°C) until adsorption equilibrium obtained for a duration of 4 h [17]. The supernatant is filtered and the equilibrium concentration (Ceq) is determined after mineralization by flame atomic absorption using a Philips type PU 900.

*Physico-chemical analysis of the rejection quality:

PH: Measured using a pH meter ORION RESEARCH type and a combined glass electrode.

Electrical conductivity (CE): measured using a conductivity type ORION RESEARCH mod 101 and a 1 cm cell.

Chemical oxygen demand (DCO): Determined according to AFNOR T90-101. The principle consists in oxidizing the organic matter contained in wastewater by an excess of silver sulfate.

Suspended solids (MES): Determined according to standard AFNOR T90-105, the filtering is done on Whatman paper (0.45 mm). The drying of the already weighed filter is done at 105°C for 1 h and weighed.

Dissolved oxygen (OD): Determined using a pulse oximeter, the assay is performed directly by immersing the electrodes in water for analysis. The oxygen reduction at the cathode generates a proportional current to the partial pressure of oxygen in the cell.

3. Characterization and treatment of global rejection by adsorption on raw chitin

3.1 Physical-chemical quality of wastewater to be treated

3.1.1 Physical-chemical quality

In order to determine the degree of pollution caused by this unit, we were brought during this work to study the physical–chemical quality of rejection to be treated. We will translate the physicochemical parameters evolution during a period of study 20 days, study period from 7 to 9-08 to 3-10-08 in **Figure 1** below.

*PH: It is an important physiological parameter which influences the development of numerous microorganisms [21] as well as speciation and the solubility of heavy metals. During followed laborers, the pH value of this discharge (**Figure 2b**)

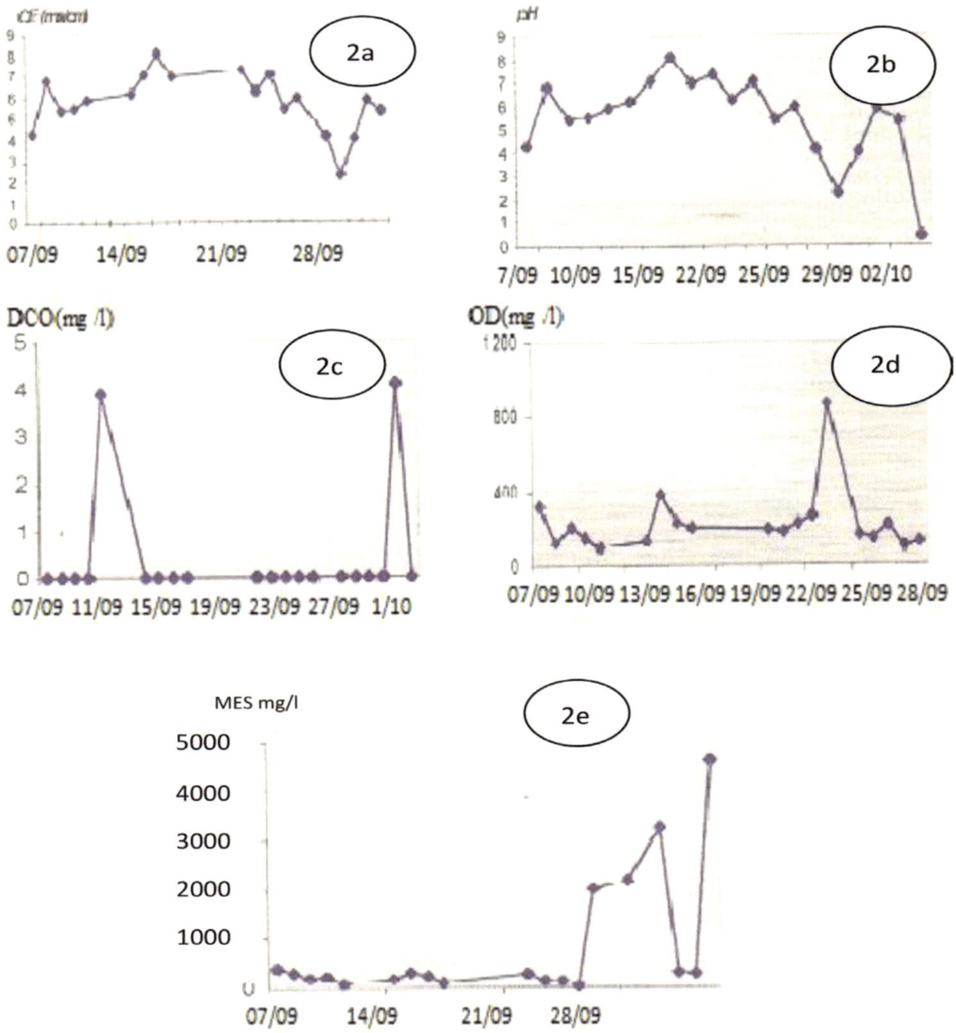

Figure 1.
Temporary follow-up of the physico-chemical parameters of the global rejection of surface treatment unity [(2a) CE, (2b) pH, (2c) DCO, (2d) OD, and (2e) MES].

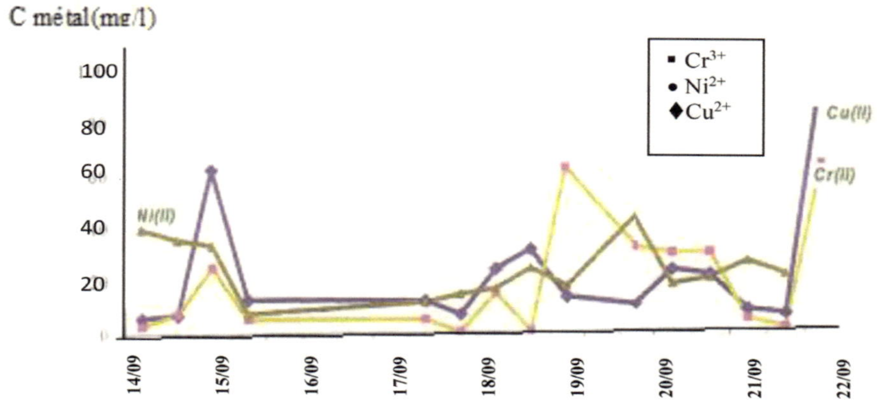

Figure 2.
Daily evolution of the heavy metals concentration in the discharge.

Sampling	Drain
22-9-08	Bath drain depassivation sulfuric
26-9-08	Bath chemical degreasing
29-9-08	Baths depassivation hydrochloric
30-9-08	Bath depassivation sulfuric Bath depassivation hydrochloric Bath anodic degreasing
3-10-08	Bath Accident at work, breaking of a pickling bath in full swing

Table 1.
Dates and the bath affected by drain.

fluctuate generally between 5 and 7, it is the optimal pH for the treatment by the raw chitin [22], except some exceptional cases such the case samples of days 22, 26, 29, 30. This is due to oil changes that have occurred in these samples (**Table 1**). Minimum values are recorded in these samples.

*CE: The electrical conductivity varies generally between 3 and 4 for all samples except that of the 3-10-98, which is very important. This value is effectively due to break of baths cleaning. The important values are recorded in the sampling days 22, 26 and 30. This drain of the baths shows that the degreasing depassivation waters of the baths are enormously salted. This salinity is essentially due to the high chloride concentration and to high acidity. The average value is of the order of 3.96 ms/cm.

*MES: The MES content confirms the statements made above. Levies where are oil changes occurring are charged by the MES. Indeed, the MES concentration is very high up to a maximum of 4.59 g/l and so exceed those generally encountered in domestic wastewater [23]. This result can be explained by the release of metallic waste and the solid deposits which accumulate at the bottom of a bath.

DCO: It present contents in variable organic matters from 96 to 3240 mg/l, but in general the value is situated near 200 mg/l, they are lower in standards dictated by the limits values of the indirect discharge [24].

OD: **Figure 2d** of OD brings to light an almost permanent state of anaerobiosis. The invalid contents of the oxygen in this discharge are due to the biological activity and to the absence of contributions in oxygen. A deficiency of this element in such effluent can have serious implications for their treatment, fermentation, release of smell, etc. This characterization shows that the wastewater of the unit can be considered relatively stable if we eliminate the variations dictated by the draining. In other words the effluent can be easily handled if we avoid the draining or we get back them in the other pipe to treat them to part and thus insure a continuous treatment of the global discharge of the unit (**Table 1**).

3.1.2 Metallic pollution analysis

The global discharge contains numerous metals that cannot be separated. We focused our study on three metals Cu, Cr and Ni. The results of the analysis of heavy metals in the effluent are illustrated by the **Figure 2**.

Cu: The levels of Cu^{2+} vary from 5.12 mg/l to 97.76 mg/l (**Figure 2**), the temporary fluctuations in this element are much more pronounced. The registered minimal value exceeds 5 times the PVL (1 mg /l). The contents of Cu^{2+} achieve in average 22.21 mg/l. This is due to currents after plating rinses.

Cr: The concentrations of chromium fluctuate between a minimal value of 0.11 mg/l and a maximal value of 63.86 mg/l (**Figure 2**). The most values exceed the PVL of Cr (2 mg/l).

Ni: as far as the nickel is concerned, the registered concentrations are enormously important and far from being in compliance with national standards. The maximal content is registered the takings of 28-9-08.

From these results, we can identify the following points:

1. Among the three metals Cu, Ni and Cr no one presents normal means in comparison with the project national standards (PVL) and the international standards (FAO, EQO, etc.) This is due to the fact that, in general, the step of dead rinsing is often exceeded.

2. The average grade is the highest registered in the case of Ni (**Figure 2**). This is due on the one hand to the fact that the nicklage is in the most part of chains and on the other hand to the fact that the standards of the bath Ni (300 mg/l) are the most raised with regard to the other one baths of plating.

3. The temporary fluctuations in heavy metal contents are essentially explained by the following client commands. The bath can work hundreds of parts by hours, consequently current rinsing will strongly by loaded. The minimal values are recorded in the case of Cr.

It appears from these results that the effluent of this unit presents a big risk on the receiving environment; this is by accumulation along the food chain of the enormous quantities of rejected heavy metals [19]. During the study of the impact of the metallic pollution on the Casablanca coast, it showed that the dosage of metallic elements in the biological compartment crab *Eriphia spinifrons* of Fe, Cr, Pb, and Cd in bivalves *Mytilus* sp. *Mactra* and *Corallina* are rather high, which indicates a possible threat of the health because of the consumption of these mollusks (**Table 2**).

3.2 Treatment test for global rejection

The results for the Cu^{2+} are summarized in **Table 2**.

Figure 3 indicates the changes of removal rates of heavy metals with the change of added quantity of raw chitin. For every mass, the raw chitin reduces the residual amount of Cu^{2+} in all treated samples even if the used quantity is weak, quoting the example of 250 mg.

Ccre				Ccra			
Test 1 Co = 0.78 mg/l		Test 2 Co = 1.31 mg/l		Test 1 Co = 0.78 mg/l		Test 2 Co = 1.31 mg/l	
Weights (mg)	Ceq (mg/l)	Weights (mg)	Ceq (mg/l)	Weights (mg)	Ceq (mg/l)	Weights (mg)	Ceq (mg/l)
30	0.45	30	1.05	50	0.17	50	0.57
60	0.32	60	1.00	100	0.16	100	0.42
90	0.30	90	0.82	150	0.13	150	0.89
120	0.28	120	0.76	200	0.12	200	0.27
150	0.25	150	0.67	250	0.07	250	0.19
180	0.15	180	0.57	300	0.05		

Table 2.
Evolution of the equilibrium concentration according to the dose of the material to remove.

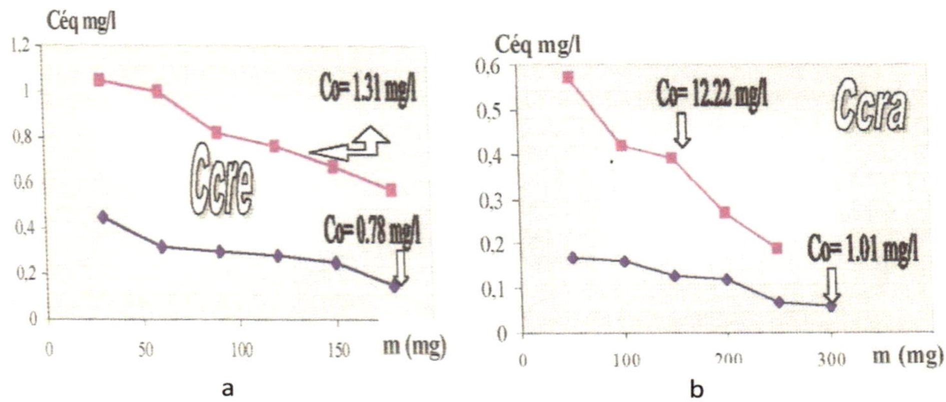

Figure 3.
(a) Variation of the Cu concentration as a function of added mass of Ccre. (b) Variation of the Cu concentration as a function of added mass of Ccra.

By calculating the average percentage removal of Cu for both materials (**Table 3**), we noticed that the highest percentages of elimination are marked for water treated by the Ccra. Whereas the percentages relative to Ccre are weaker than what is in agreement with [16].

For a given sample, when the mass of material increases, the percentage of reduction increases slightly. The variation of the percentage change can be considered constant from 250 mg (**Table 3**).

	Ccre		Ccra	
	Test 1	Test 2	Test 1	Test 2
Co mg/l	0.78	1.31	1.01	12.22
Ceq mg/l	0.25	0.67	0.13	0.39
%	67.95	48.85	87.13	96.80

Table 3.
Average removal percentages evolution of Cu for both materials.

Ni: To estimate the efficiency of the material for the treatment this wastewater, we followed the evolution of the residual concentration of the Ni according to the various injected doses. The optimal dose of the material is chosen according to the quality wished by the water treaty. It is generally obtained when the ratio of M^{2+} equilibrium/M^{2+} original becomes little bit constant.

Figure 4 shows that the residual amount of Ni after adsorption on the Ccre decreases gradually by increasing the dose of the material. For the Ccra the concentration in the equilibrium after treatment by 50 mg does not differ any more from that stayed after treatment by 150 or 200 mg. This results shows that the Ni shows an affinity important for Ccra. This is at the middle in evidence by the efficiencies on elimination which are maximums for Ccra.

Cr: The results of the dosage of the chromium before and after adsorption are included in **Table 4** below.

From this table, the removal efficiency increases as the Ccre metal concentration decreases. This agrees well with the isothermal studies [16], even if the physico-chemical quality of the water differs. We should note also that all the initial concentrations exceed the PVL fixed to 0.2 mg/l (**Table 5**). After treatment by the Ccra and by Ccre, the concentrations become lower than the standards. The hexavalent chromium is weakly eliminated by Ccre and

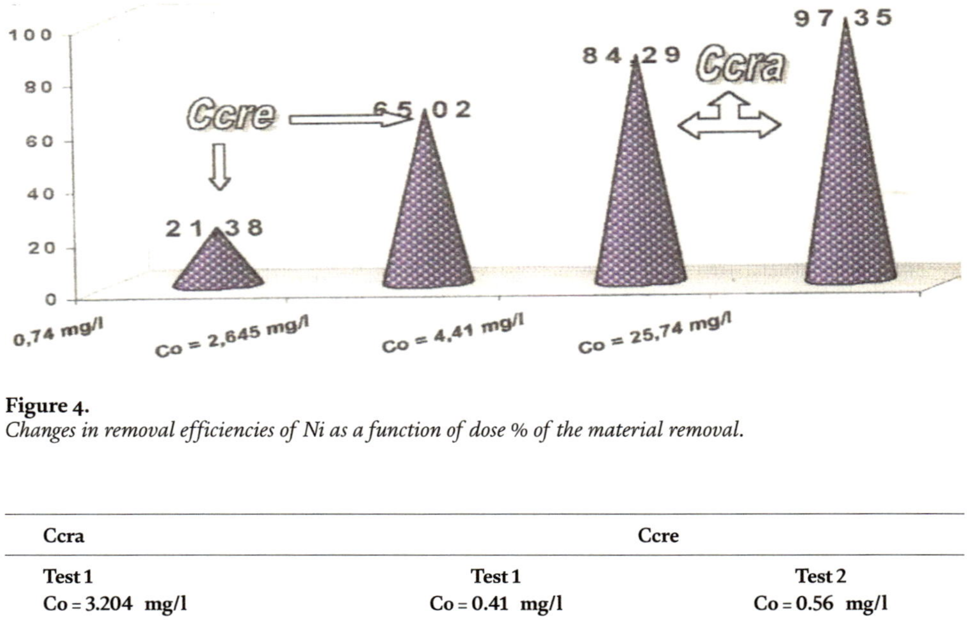

Figure 4.
Changes in removal efficiencies of Ni as a function of dose % of the material removal.

Ccra		Ccre			
Test 1 Co = 3.204 mg/l		Test 1 Co = 0.41 mg/l		Test 2 Co = 0.56 mg/l	
Weights (mg)	Ceq (mg/l)	Weights (mg)	Ceq (mg/l)	Weights (mg)	Ceq (mg/l)
	0.036	30	0.096	30	0.051
100	0.044	60	0.024	60	0.100
150	0.016	90	0.002	90	0.060
200	0.011	120	0.018	120	0.087
250	0.053	150	0.014	150	0.056
300	0.020	180	0.019	180	0.033

Table 4.
Evolution of the concentration of the Cr^{6+} according to the dose of the material.

	Ccre	Ccre	Ccra
Co (mg/l)	0,411	0,564	3,204
Ceq (mg/l)	0,024	0,100	0,044
PVL (mg/l)	0,2	0,2	0,2

Table 5.
Evolution of the concentration equilibrium Ceq according to the initial concentration C_o for a dose of 0.6 mg/l.

by Ccra [14, 16], yet in this discharge, the elimination of chromium is very important by both sources of the chitin.

4. Characterization and treatment of current rinsing by adsorption on raw chitin

4.1 Physical-chemical characterization

4.1.1 Physical-chemical

We have studied the characterization and the treatment of the rejects of the metallization baths of metals Cu, Zn, Cr and Ni. The physico-chemical rinses aware of four baths metallization (Cu, Zn, Cr and Ni) has been grouped by the following table:

pH: Because of degreasing, etching and galvanic deposition, we worked with solutions of different types of reactions. We have to avoid absolutely the training of the slightest traces of a solution in what is next. In addition, we must ensure that no residual solution in the emptiness or back of the room, because this residue would affect extremely, adversely the adhesion of a plating. According all to the possibilities the rinsing must be done in that is rinses current water which is characterized by a neutral pH, from **Table 6**, the recorded pH is more neutral. The pH of the rinsing current Cr is acidic. All rinses Cr acquire the characteristics of a flushing death. The same for the flushing power of Cu, since it works with an alkaline bath, the pH of the rinse is relatively high; it reached a maximum 8.33 for the collection of 11-9-08. In the case of current rinsing bath of Zn is acid so the pH values below 7. They reach 6.6 by the same observation was recorded in the case of power flushing Ni. Note that for the same type of rinsing, the pH does not change significantly from one chain to another.

CE: Electrical conductivity is the lowest recorded in the case of the Cu current rinsing, while the highest values are recorded for the zinc rinses (**Table 6**).

MES: rinses have higher levels of MES ranging from 162 to 740 mg/l respectively for Ni and Zn. MES in the rinses is much smaller than the global rejection [14]. The rinses of zinc are the most loaded (**Figures 5** and **6**).

DCO: Highest values of DCO are recorded in the case of rinsing the zinc in the chain II, while for the rinsing of Ni, Cu and Cr, they do not exceed by 400 mg/l, that value is less the limit values (PVL).

OD: it is large fluctuations, water bodies are generally well oxygenated (**Table 6**). Maximum values up to 17.8 mg/l are noted in the current Cu rinses. This contribution is due to the complete absence of a biological pollution in the rinse tanks. Besides all the settings in this collection are low, the DCO does not exceed 40 mg/l. For comparison the variation of average concentrations of these parameters depending on the type of metal rinsing are shown in **Figure 5** below.

	Day/ month	Chain	pH	CE (mS/cm)	MES (mg/l)	DCO (mg/l)	OD (mg/l)	SO_4^{2-} (mg/l)	Cl- (mg/l)	Ca^{2+} (mg/l)
	11/9	II	6.96	3.93	190	380	14.6	212	—	—
Ni^{2+}	15/9	II	6.81	3.66	162	—	16.6	348	537	—
	23/9	II	6.56	3.73	134	300	12	200	521	—
	3/10	II	6.72	4.00	—	340	10.0	64.8	507	736.2
	11/9	I	6.04	49.50	525	2960	4.9	631	—	688
Zn^{2+}	15/9	I	6.07	43.75	500	3920	6.41	556	—	—
	23/9	I	6.04	63.79	740	2800	5	500	—	785.8
	3/10	I	6.08	58.0	715	3760	6.6	424.8	724	883.4
	11/9	II	3.02	3.93	492	258	10.2	—		—
Cr^{6+}	15/9	IV	3.21	8.16	234	315	9.2	202	—	—
	29/9	II	4.12	3.05	138	418	5	353	—	—
	3/10	IV	2.5	17. 5	74	269	121.4	50.4	—	272
	11/9	II	8.33	3.04	88	480	8.5	—	597	112.1
Cu^{2+}	23/9	IV	7.51	2.13	138	40	17.8	37.9	239	192.1
	3/10	I	8.19	2.24	628	400	12.01	13.9	269	160.0
	3/10	I	7.97	3.74	190	240	14.7	—	726	288.1

Table 6.
Quality physical-chemical of rinses for the treatment.

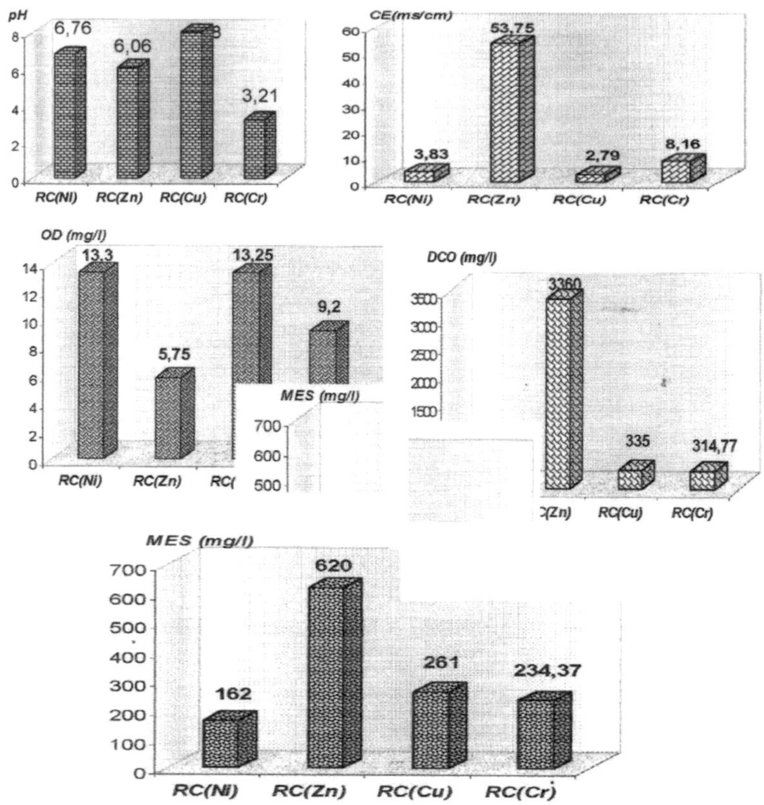

Figure 5.
Evolution of the physicochemical parameters pH, EC, MES, COD, DO according to the type of current rinsing [RC(Ni), RC (Zn), RC(Cu) and RC(Cr)].

4.1.2 Metal pollution

In order to assess metal pollution, we tried to assay the ion currents located in the rinsing after plating by Cu, Zn, Cr and Ni (**Figure 6**).

In the studied unit, the static rinse is used primarily to recover and recycle metal in the plating baths, but on the other hand, the current simple rinsing aims to clean the pieces. The determination of heavy metals in these rinses (**Table 7**) shows that the levels of metal ions are very high. They are generally higher than those found in the global rejection. Example, the levels of Cu^{2+} reach an average 22.21 mg/l [14].

In fact, some bath rinses as Ni and Cr are highly concentrated in the rinse tanks, there are many colored water after rinsing the metal: green for Ni, Cu and blue to yellow for Cr. These waters differ slightly from the dead rinses. These huge losses of metal ions are due to inadequate drainage time pieces and often exceeded the stage of rinsing death. For the case of Cu the maximum values are found in the flushing of the IV chain, while the levels of the chain I are relatively low, this is due to the number of work pieces and the capacity of each channel.

For the case of Ni the values fluctuate between 15 and 34 mg/l with an average of 24.51 mg/l. In the case of Zn, the maximum values are recorded in samples of 11-9 and 3-10.

It should be noted also that in the current flush of a metal there are traces of other metals, rinsing the Ni the chain of 3-10 II also contains Cu and Zn, Cu rinsing contains Cr, Cr rinse contains Ni and Cu (**Table 8**).

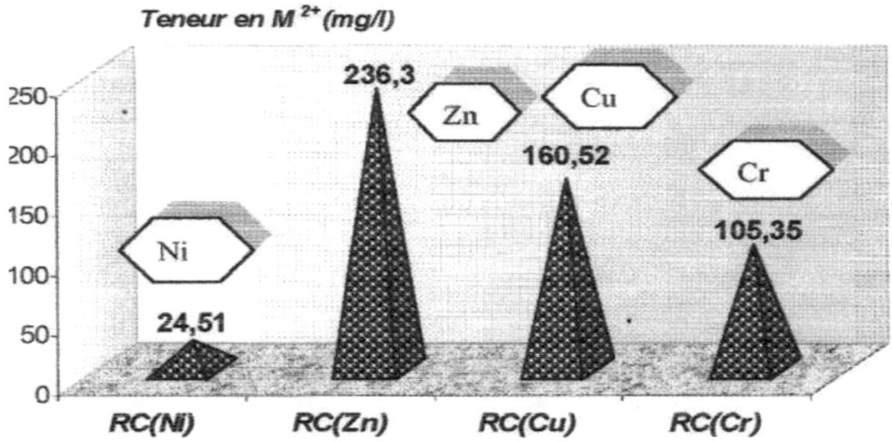

Figure 6.
Average levels of metal ions in the rinse currents.

	Nickel	Copper	Chrome	Zinc
11/9	34.21(II)	296.5(IV)	211.7(II)	201.1(I)
15/9	29.31 (II)	—	20.13(IV)	32.4(I)
23/9	19.02(II)	20.70 (I)	—	41.6(I)
29/9	—	—	0.96(II)	—
3/10	15.05(II)	256.6(IV)	—	670.1(I)
3/10	—	68.3(I)	188.6(IV)	—

Table 7.
Evolution of metal ion concentrations (mg/l) according to the type of rinsing in different chains.

	Ni	Cr	Zn	Cu
Current rinse of Ni	15.05	0.59	1.93	0
Current rinse of Cr	3.87	188.6	0	12.42
Current rinse of Cu	0	1.17	0	68.3
Current rinse of Zn	0	0	670.11	4.73

Table 8.
Concentration of metal ions in the rinse of metal in mg/l of sample 3/10.

4.2 Rinsing water treatment test common to plating baths

The current water rinses are treated the same way that releases overall. We studied the case of Cu, Ni, Cr and Zn. For each metal we have tried to work on two different samples. The results of this study are summarized in the **Table 9** below.

The expression of the calculation of the % of adsorption is 100 * C0-Ceq/C0.

Cu: The table shows that the removal efficiency increases as the mass of material increases especially for the rejection diluted. In general, treatment of rinse water is more effective than the treatment of Cu in the global rejection [6–8]. Indeed, the removal percentages do not exceed more than 63% in the global rejection while for flushing streams they reach 98.08%.

Ni: In the case of Ni, we see that the residual concentration decreased from 1.5 to 1.2 mg/l with a percentage reduction is very small compared to the percentage reductions in the case of total rejection [6], we can say that the presence of other metals does nothing, but increasing the removal of the Ni.

Cr: the removal yields is relatively low compared to those found in the overall rejection, it does not exceed 28% for the diluted rinsing and 25% for the concentrated rinsing.

		Weights of adsorbant (mg/l)	50	100	150	200
	1	C (mg/l)	1.99	2.69	3.99	0.43
		% of adsorption	92.23	89.49	84.45	90.17
Cu	2	C (mg/l)	0.59	0.54	0.31	0.12
		% of adsorption	71.64	74.06	84.98	94.39
	1	C (mg/l)	165.7	155.9	141.1	149.6
		% of adsorption	12.14	17.34	25.18	20.68
Cr	2	C (mg/l)	15.39	15.43	16.15	15.10
		% of adsorption	27.3	27.11	23.71	28.67
	1	C (mg/l)	63.66	—	16.52	6.54
		% of adsorption	4.99	—	16.83	90.23
Zn	2	C (mg/l)	0.290	0.051	0.039	0.033
		% of adsorption	—	—	—	—
Ni	1	C (mg/l)	1.2	—	—	—
		% of adsorption	66.27	—	—	—

Table 9.
Changes in percentage of the adsorption of metal ions of the current rinses depending on the dose of added material.

Zn: to achieve a removal efficiency of 90.23%, it is necessary to introduce a mass of 200 mg, Similar to Cu, the effect of the dose of the material added to the percentage reduction of micro metal is very pronounced for high concentrations of metals. This result confirms the results found by Boukhlifi et al. [17, 30, 31]. By comparing the removal efficiency of the four metals, it appears that the strongest removal is marked for the case of Cu [16].

5. Study flow of pollution

A chain consists of a set of tanks whose general functions are: surface preparation, processing and finishing of the piece part in question. Each tank is defined by three characteristics:

- The mode of treatment (pretreatment, metal deposition, stripping ...)
- The type of chosen treatment (e.g. a metal deposit: silver, chrome plating, copper plating, zinc plating, etc.)
- The operating conditions (for example, a zinc plating: alkaline non-cyanide, acid fluroborate not ...) including different chemicals concentrations, fluid flow, the rate of production, etc.

The unit of surface treatment under study is composed of five chains. Each channel is determined by the succession of tanks. The average capacity of baths varies between 950 l and 1710 l, but most of the bath has a volume of 1440 l, the majority of the baths is powered water wells except metal plating baths and degreasing baths, which are filled with drinking water, bathing water that is recycled in metal plating baths of rinses dead. Water supply wells are often for an hour a day, while drinking water is draining after a bath. Discharges baths is collected through pipes that lead to the aerated sewage. All discharges are evacuated in the rough, but the rejection of the flushing stream which is recycled zinc. We were interested in flushing power of Ni and Cr; we followed up daily flow rates of 7/9 to 3/10 in **Figures 7** and **8**.

We found that the high flow rates were recorded for Cr and Ni, the rate of flushing power of Cr can be up to 600 l/h and the neither flushing current reached a maximum of 841 l/h.

- The flow rates of Ni in the chain will fluctuate between 260 and 390 l/h, they will reach 389.21 l/h, a value which represents the 1/3 of the bath of treatment, that is to say that every day, the third of the rinsing bath is changed with a capacity of 1440 l. This during the fifth of the rinsing bath of Cr is changed. These waters are evacuated; this appears from the color of releases. Rejection of Cr is yellow while the rejection of Ni is green.

The Ni current rinsing flow varies from one channel to another varies from the average flow 389 l/h for channel II 738.8 l/h for the string I.

- The temporal variation of the flow does not follow a given order; it varies from day to day depending on water supplies that are directly related to the availability of water in the well. - We tried to compare these rates with those of other baths as bath chemical degreasing and pickling bath. The flow of degreasing bath is lower and messy, while the stripping is relatively constant and is around 400 l/h.

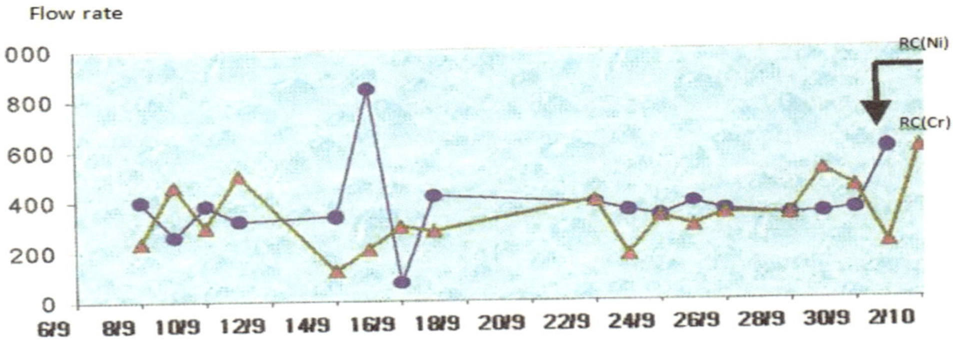

Figure 7.
Monitoring rinsing flow flows as a function of time.

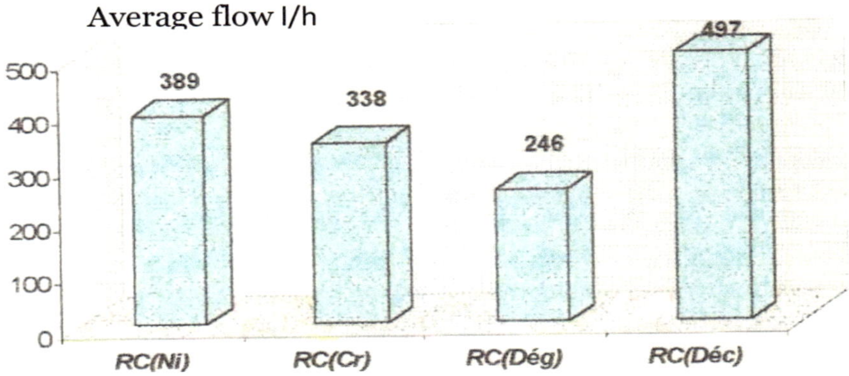

Figure 8.
Average flow rates of rinses for different metals.

From this study, it follows that the rates baths variation is random. This makes the adaptation of a method for processing or recycling of waste water very difficult. In conclusion, we must control the flow based on the minimum and maximum values recorded by installing a storage buffer.

6. Description of procedure for treatment

Handling treatment facilities are numerous, these some examples of treatment courses of this type of release:

Physical–chemical treatment is to continuously purify the various effluents as when they are produced to reject them in permanent rates.

The process is a combination of the following basic treatments:

- manure storage
- basic treatment
- additional treatment
- Final Control
- Physical–chemical batch which is to purify the effluent by tarpaulin, it is to say by successive and constant volume in the same reactor of suitable sequences [16].

- In general the combination of elementary processing is defined by the nature of the effluent to be treated. According to preliminary studies we conducted, only the rinse water can be recycled after treatment with raw chitin in particular rinses common metallization baths [25–28]. We therefore propose a processing rinses followed by a comprehensive treatment of rejection. The design of items of treatment plant is highly dependent on flow rates of each rinse that helps to estimate the reaction time stays easily adaptable.

- Reactors as a column filled with the material well-conditioned. Water to be treated flows from the bottom up by the effect of differences in level and then flows into the rinsing baths continuous overflow. Thus the heavy metal content in the total discharge will be reduced.

For the global rejection we prefer a continuous physical–chemical treatment. This principle (continuous) is used in preference to batch treatment by tarpaulin, where daily volumes in excess treat tens of m^3 even if the discontinuity of the process provides some security and facilitates the monitoring and treatment by visual inspection of each analytical or steps prior to discharge. It allows, if necessary, replacement of the entire process or just step failed. This treatment considered contains the following five basic treatments [27–31]:

1. Effluent Storage: The storage provides the interface between production facilities and the treatment plant so it performs the following functions:

    - A buffer role with regard to changes in volume and mass charges from workshops, and a reserve sufficient to accommodate the effluent during the treatment.

    - A perfect separation between the different components of the effluent.

    - A minimum hydraulic capacity, allowing the complete filling of the reactor treatment at the launch of the operation with the more regular flow of possible mergers with minimal variations.

The storage to consider is divided into three types of storage:

- Storage backup's role is to provide storage during an accident or during an overload of other storage.

- Storage diluted normally receives all current daily rinsing of all pre-treatments and treatments.

- Storage is focused especially on the garbage baths surface preparation or special cases.

The storage times are equipped with means for adjustment and control of flow injection in the treatment reactors.

2. liquid–solid separation step that gets rid of the effluent MES and some insoluble elements [31]; it is installed mainly concentrated after storage. The liquid–solid separation of sludge is achieved by settling, both phases are separately drained.

3. Neutralization, this phase is on the middle that is, the characterization studies have shown that most of the pH is close to 5.99 with an average 6.01 except a few exceptional cases. From then this step will be in most of the day exceeded even in the case of the neutralization of dilute effluents at the exit of treatment, discharge will be clarified concentrated alkaline after treatment and then joined the others in terms of diluted effluent neutralization.

4. Specific Treatment: The treatment methods are defined in terms of chemical elements in the flue. In most cases, the courses include.

- One or more stages of detoxification.
- A phase in dissolution-precipitation.
- A flocculation phase alternating.

But all these steps, we offer the reactors filled with raw chitin from shrimp processing is done in series with at least two reactors. Power through the discharge is from the bottom up to ensure the best conditions of contact. Similarly we must always place two reactors relief and it is also a book containing alternating chain also at least two presidents to keep the operation continues.

5-Treatment of sludge generated once the carrier material is saturated, the sludge is then pumped out regardless of the waters. This sludge can be regenerated by an acid or recycled in building materials [32], such transactions may take place after dehydration on a filter press which the design is based on the mass flows of raw material and insolubilized in the reactor.

7. Conclusion

In conclusion, and since the coast of Casablanca city is subjected to numerous anthropologic attacks, In front of such a situation, it is indispensable that every industrial unity must be equipped with a wastewater system treatment, to protect the environment and its resources. The treatment of the global discharge shows that there is competitive adsorption between metals. The nickel has more affinity toward crab raw chitin. The waste water treatment by raw chitin from shrimp entrains a decrease of heavy metals contents in the global rejection. The efficiencies on elimination are important and excess 99% for some metals. The use of the raw chitin shows itself of big potential for the treatment of the liquid discharges of the studied unity MAFER. It showed itself capable of treating heavy metals loads superior to 200 mg/l by presenting percentage removal between 90% and 97%, as in the case of Cu^{2+}.

The physic-chemical characterization shows that the currents rinses are greatly loaded with heavy metals far exceed the proposed limit values Morocco PVL. The adsorption test of heavy metals on raw chitin showed interesting results for the current rinses. Besides the rejection colored before adsorption (case of Ni, Cu, and Cr) becomes colorless after treatment with raw chitin. The study of the rinses currents treatment by raw chitin, we allowed the description of a treatment facility based on chitin. Consequently the adsorbing support raw chitin must be added in the treatment stations of the effluent of the units of surface treatment to eliminate the metallic pollution.

Author details

Boukhlifi Fatima
Materials and Applied Catalysis Equip (MCA), Laboratory of Chemistry and Biology Applied to the Environment, Faculty of Science, My Ismail University, Meknes, Morocco

*Address all correspondence to: boukhlifi1@yahoo.fr

References

[1] Kurniawan TA, Chan GYS, Lo W, Babel S. Physico-chemical treatment techniques for wastewater laden with heavy metals. Chemical Engineering Journal. 2006;**118**:83-98

[2] Charerntanyarak L. Heavy metals removalby chemical coagulation and precipitation. Water Science and Technology. 1999;**39**(10/11):135-138

[3] Ying X, Fang Z. Experimental research on heavy metal wastewater treatment with dipropyl dithiophosphate. Journal of Hazardous Materials B. 2006;**137**:1636-1642

[4] Li YJ, Zeng XP, Lius YF, Yan SS, Hu ZH, Ni Y-M. Study on the treatment of copper electroplating wastewater by chemical trapping and flocculation. Separation and Purification Technology. 2003;**31**:91-95

[5] Lazaridis NK, Matis KA, Webb M. Flottation of metal –loaded clay anion exchangers. Part I: The case of chromate. Chemosphere. 2001;**42**:373-378

[6] Doyle FM, Liu ZD. The effect of triethylenetetraamine on the ion flotation of Cu2+ and Ni2+. Journal of Colloid and Interface Science. 2003;**258**:396-403

[7] Belkacem M, Khodir M, Sekki A. Treatment characteristics of textile wastewater and removal of heavy metals using the elctroflotation technique. Desalination. 2008;**228**:245-254

[8] Alvarez-ayuso E, Garcia-Sanchez A, Querol X. Purification of metal electroplating wastewater using zeolites. Water Research. 2003;**37**(20):4855-4862

[9] Inglezakis VJ, Zorpas AA, Loizidou MD, Grigoropoulou HP. The effect of competitive cations and anions on ion exchange of heavy metals. Separation and Purification Technology. 2005;**46**:202-207

[10] Casas I, Miralles N, Sastre A, Aguilar M. Extraction of cadmium by organophosphorous compounds. Polyhedrom. 1986;**5**(12):2039-2045

[11] Juang RS, Shiau RC. Metal removal from aqueous solutions using chitosan-enhanced membrane filtration. Journal of Membrane Science. 2000;**165**(200):159-167

[12] Katsou E, Malamis S, Haralambous KJ. Industrial wastewater pre-treatment for heavy metal reduction by employing a sorbent –assisted ultrafiltration system. Chemosphere. 2011;**82**:554-564

[13] Ku Y, Peters R-W. The use of activated carbon as a polishing step for treatment of industrial plating wastewater. In: Proc. Nat. Conf. Envron. Eng. 1982

[14] Boukhlifi F, El Akili C, Moussout H, Benzakour A, Ahlafi H. Treatment of global rejection of electroplating industry by raw chitin. International Journal of Applied Environmental Sciences. 2013;**8**:13-23

[15] Boukhlifi F. Purification of the waste water of the galvanoplasty industry by natural adsorbents: Current rinsing, GHAZES N. 2; 2018

[16] Boukhlifi F, et Bencheikh A. Etude de l'adsorption compétitive des métaux lourds sur la chitine brute: Application aux eaux usées d'une industrie chimique. LA Tribune de l'eau, No 611/3, Vol. 55; 2001, pp. 37-43

[17] Boukhlifi F, Bencheikh A. Characterization of natural biosorbents used for the depollution of waste water. Annales de Chimie Science des Materiaux. 2000;**256**:153-160

[18] Demirbas A. Heavy metal adsorption on to agro-based

waste materials: A review. Journal of Hazardous Materials. 2008;**157**:220-229

[19] Sik Ok Y, Yang JE, Zhang YS, Kim SJ, Chung DY. Heavy metal adsorption by a formulated zeolite-Portland cement mixture. Journal of Hazardous Materials. 2007;**147**:91-96

[20] Ayres J-L, Fedkiw P-S. Abatement of heavy metals industrial effluents by a catalyzed electrochemical removal scheme, Report No. 207, water Resources Research Institute of the University of North Carolina, North Carolina State University; 1989

[21] Schoeman J-J, Vanstaden J-F, Saayman H-M, Vorster W-A. Evaluation of reverse osmosis for electroplating effluent treatment. Water Science and Technology. 1992;**25**(10):79-93

[22] Zaloum R, Drolet R, Ward R, Levasseur C. Application du concept effluent zéro à l'usine Uniforêt Inc. de Port Cartier, Fiche d'un projet de développement et de démonstration technologique réalisée par la compagnie Uniforêt; 1998

[23] Khayati A, Messafi A. Traitement des rejets liquides émanant des industries de traitement de surface et leur réutilisation dans les circuits des chaines selon le principe « rejet zero ». Desalination. 2004;**167**:87-99

[24] Sigg L. Surface chemical aspects of the distribution and fate of metal ions in lakes. In: Aquatic Surface Chemistry, Chemical Processes at the Particle-Water Interface. New York: Wiley-Interscience; 1987

[25] Bechac JP, Boutin P, Mercier B, Nuer P. Traitement des eaux usées, ed. Eyrolles, Vol. 6-8; 1984. pp. 104-145

[26] Agence de l'eau, Rhône Méditerranée Corse, SITS, Traitement de surface, Epuration des eaux; 1998

[27] Boukhlifi F, Bencheikh A, Ahlafi H. Characterisation and adsorption propriety of chitin toward copper cu2+. Physical and Chemical News. 2011;**58**:67-72

[28] Boukhlifi F. Etude de la rétention des micro-polluants métalliques (Pb, Cd, Cu et Zn) sur de nouveaux matériaux biosorbants: Essais d'épuration d'effluents industriels liquides [thèse de doctorat]. 2000

[29] Moussout H, Ahlafi H, slimani MS, Boukhlifi F, Daou I. Bentonite/chitosan biocompositeϖ as an adsorbent for hexavalent chromium from aqueous solutions. Journal of Advanced Chemistry. 2014;**10**:2786-2795

[30] Slimani MS, Ahlafi H, Moussout H, Boukhlifi F, Zegaoui O. Adsorption of hexavalentϖ chromium and phenol onto bentonite modified with HexaDecylTriMethylAmmonium bromide (HDTMABr). Journal of Advanced Chemistry. 2014;**8**:1602-1611

[31] Caughman, Jr CR, Houston, Tex. sludge filtration system for separating the sludge liquids, United States Patent (19) 11, Patent Number: 5,595,654; 1995

[32] Manal N, Samdi A, Elabassi K, Gomina et M, Moussa R. Recycling of industrial wastes, phosphogypsum and fly ash. In: Building Materials, IInd International Seminar INVACO – Innovation & Valorization in Civil Engineering & Construction Materials , MATEC Web of Conferences. Vol. 2. 2012

Chapter 8

Analysis of Mechanical Properties of Austempered Ductile Iron Weld Joints Using Developed Electrode

Tapan Sarkar

Abstract

In the present investigation mechanical properties of austempered ductile iron (ADI) joints have been studied using two developed electrodes containing with and without Ce content and co-related between microstructure and mechanical properties. Austenitization was done at 900°C for 2 h and austempering at 300 and 350°C for 1.5, 2 and 2.5 h holding time. At 300°C microstructure shows needle shaped bainitic ferrite with lower amount of retained austenite indicate higher hardness and lower impact toughness. However, at 350°C shows feathery shaped bainitic ferrite with higher amount of retained austenite to demonstrate lower hardness and higher impact value. Both the welded joints at both austempering conditions tensile samples broke from the base metal indicates 100% joint efficiency. Fatigue life was varied with varying the austempering temperature and shows higher fatigue life at 350°C austempering temperature presence of higher amount of retained austenite and finer the bainitic ferrite size with smaller graphite nodules. Ce in weld metals to refine the microstructure and shows higher impact toughness and fatigue strength with lower hardness value at both austempering temperatures.

Keywords: heat treatment, microstructure, tensile test, Charpy impact, high cycle fatigue

1. Introduction

Austempered ductile iron (ADI) is a new family member of engineering materials. It has recently received significant attention owing to the excellent combination of mechanical properties such as high strength together with good ductility, good wear resistance, and higher fatigue strength to make the material as a successful substitute for forge steels or aluminum alloys [1–4]. The remarkable properties of the ADI are attributed with the unique microstructural constituents such as bainitic ferrite and high carbon enriched retained austenite.

The low production cost and production advantage of ADI, it has been used in many structural applications and many critical parts of automobiles such as crankshaft, steering knuckles, hypoid rear axle gears, camshafts and disk-brake calipers etc. in which fatigue resistance is an important requirement [5, 6].

ADI shows higher fatigue life than as-cast ductile iron (DI) and determined by generating stress-life (S-N) curves [7]. The fatigue life of ADI strongly depends on the austempering temperature, austempering holding time, austempering kinetics,

IntechOpen

the amount of retained austenite, shape and size of bainitic ferrite and graphite nodules [1, 3].

The chemical composition of ADI is similar to conventional DI. However, some alloying elements such as Ni, Cu and Mo are usually added to increase the austemperability, to delay the austenite decomposition [8, 9] to pearlite and ferrite upon cooling. DI converts to ADI with the help of two steps isothermal heat treatment process. Austenitization was done at 850–910°C for 30 min–2 h holding time and austempering at 250–450°C for 5 min–4 h holding time and finally cooled at room temperature in the open air [10]. The total heat treatment process depends on the chemical composition and thickness of the as-cast DI. Austempering temperature and holding time shows significant effects on the microstructure and mechanical properties of ADI. During austempering two-stage reactions have been done, at the initial stage (stage I) austenite (residual austenite) decomposed into bainitic ferrite and high carbon austenite (retained austenite). Increasing holding time the stage II reaction was start and high carbon austenite further decomposed into bainitic ferrite and carbide (ε carbide). The time periods between the two stage reactions is called the process windows, obtained optimum combination of microstructural and mechanical properties. Presence of alloying elements such as Ni, Cu and Mn to delay the austempering reaction and increased the process windows [11].

At a lower austempering temperature ADI shows needle shaped bainitic ferrite with a lower amount of retained austenite and graphite nodules, which in turns increased the tensile strength and hardness to decrease the elongation and toughness. However, at higher austempering temperature shows coarser bainitic ferrite with higher amount of retained austenite and graphite nodules, as a result to decrease the tensile strength and hardness; increased toughness that illustrate higher fatigue strength [6, 10, 12].

It is reported, the fatigue strength of ADI is not only depended on tensile strength and hardness like as steel [13]. However, in ADI the amount of retained austenite and its carbon content, size of graphite nodules, nodularity, shape and size of bainitic ferrite plays an important role in the high cycle fatigue performance and higher fatigue limit [1, 13–15]. Bahmani et al. [16] illustrate a relationship between the microstructure and fatigue strength of ADI and obtained, the fatigue strength depended on the amount of retained austenite and its carbon content. The fatigue strength was increased as increasing the amount of retained austenite and its carbon content. Graphite nodularity and its size show significant effects in fatigue life. In ADI graphite working as a shrinkage cavities. During fatigue test, micro crack was initiated around the graphite and then formed to macro crack which leads to the final failure of the sample [17]. Sofue et al. [18] reported, with increasing the graphite size to decrease the fatigue strength and the fatigue strength was optimized by decreasing the graphite nodule size.

Further, rare earth metal such as cerium has a beneficial effect on the microstructure and properties of ADI. However, the optimum rare earth content varies significantly according to different investigators. For example, researchers [19] reported that the presence of Ce content from 0.005 to 0.014% the nodularity was increased with refining the size of the nodules but further increasing Ce content up to 0.018 or 0.020%, the nodularity decreased and formed some non-spherical graphite with coarsening the nodule size to decreased the fatigue strength. Choi et al. [20] observed that DI castings with 0.3% rare earth content attributed improved graphite nodules, lower tensile strength and hardness, higher elongation to indicate the higher fatigue strength than that of DI castings without rare earth.

However, in spite of high strength, reasonable ductility and higher fatigue life, the application of ADI is somehow limited due to non-availability of a suitable electrode which has inhibited the joining of such high potential material.

Furthermore, many a time ductile iron castings for converting ADI need to repair welding which also demands welding consumables compatible with DI base materials. Commercially available coated electrodes for welding conventional DI include pure nickel (90–97%) [21], stainless steel and iron-nickel [22] which are first of all not suitable for converting ADI from DI weld due to poor austemperability and also not cost-effective [21, 22]. Recent authors successfully developed a coated electrode for welding DI [23] and convert to ADI after heat treatment, obtained higher tensile strength and toughness. Further improvement of the weldability of DI and welding of higher grade DI to obtain higher toughness and fatigue strength, coated electrode was developed to introduce nano-CeO_2 as a flux ingredient [24].

However, all the previous literature discussed about the fatigue strength or fatigue life only on DI or ADI base metal [1, 14, 25, 26]. But there is no such a literature to discuss about the fatigue life of ADI weld joints. Owing to the importance of the ADI especially on structural and automobile application, it is needed to find out the fatigue properties of ADI weld metal to use in commercial application.

The present work thus aims is to study the mechanical properties of ADI weld joints and co-relation between as-cast and heat treated microstructure, using two developed electrodes. Microstructural studies were done using optical microscopy (OM) and scanning electron microscopy (SEM). Phase analysis was performed utilizing XRD analysis and mechanical properties of the weld joints performed under microhardness, transverse tensile, room temperature Charpy impact and high cycle fatigue (HCF) test. Fatigue crack growth and fatigue fracture surface were investigated under SEM studies.

2. Experimental procedure

As cast DI was collected from the local foundry and used as a base metal for this experimental purpose. The details chemical composition of the as-cast DI is given in **Table 1**.

Two developed electrodes containing without and with Ce such as Trial 4 and Trial 7 selected for the experimental purpose. Among the two electrodes, Trial 4 contents with without Ce and Trial 7 contents with Ce content (0.1%). The details chemical composition of the two developed electrodes is given in **Table 2**.

Modified U groove (**Figure 1**) weld was performed on the DI base plate using both Trial 4 and Trial 7 electrodes applied shielded metal arc welding (SMAW) process. Preheat was applied at 300°C for 1 h and post weld heat treatment (PWHT) at 300°C for 1 h immediate after welding using the constant welding parameters [27, 28]. The details welding parameters are given in **Table 3** and the defect free weld was considered as per AWS (D11) [29].

The welded DI specimens are converted to ADI with the help of two steps isothermal heat treatment process. Austenitization was done at 900°C for 2 h holding time and then samples are immediately transferred to salt bath for austempering process. Austempering was done at 300 and 350°C for 1.5, 2 and 2.5 h holding time than air cooled to room temperature. Typical isothermal heat treatment cycle is shown in **Figure 2**.

Element	C	Si	Mn	S	Cr	Mg	P	Fe
Wt.%	3.60	2.92	0.22	0.019	0.028	0.041	0.01	93.16

Table 1.
Chemical composition of as-cast ductile iron.

Element wt.%	C	Mn	Si	S	P	Ni	Mo	Cu	Al	Bi	Mg	Ca	Ti	Fe	Ce
Trial 4 (without Ce)	3.08	0.40	2.60	0.006	0.039	0.5	0.19	0.24	0.62	0.03	0.004	0.015	0.09	91.40	—
Trial 7 (with Ce)	3.28	0.43	2.76	0.008	0.02	0.66	0.31	0.29	0.48	0.03	0.003	0.003	0.14	91.45	0.10

Table 2.
Chemical composition of the weld deposits using developed coated.

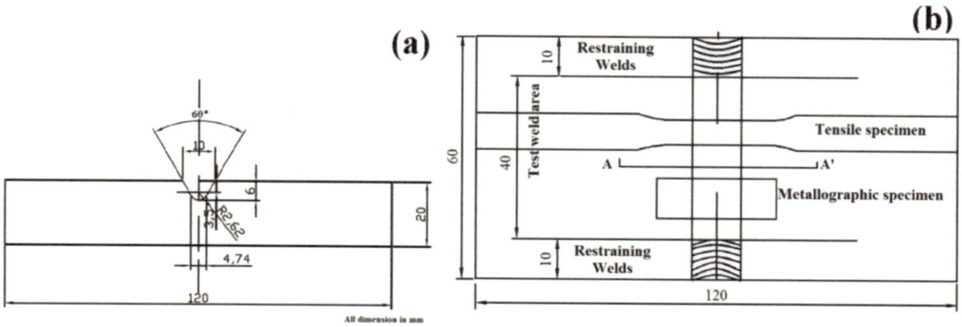

Figure 1.
(a) Modified grove design, (b) schematic view of extracting samples from weld metal.

Parameters	Unit	Value
Preheat temperature (1 h)	°C	300
PWHT (1 h)	°C	300
Welding current	A	150
Arc voltage	V	24
Welding speed	mm/s	1.70
Heat input	kJ/mm	1.58

Table 3.
Defect free weld procedure.

Metallography samples of 20 × 10 × 15 mm dimensions were cut from the large size weld specimen and removed any decarburized skin by surface grinding. Samples were mounted at room temperature, then polished systematically in silicon carbide paper and followed by cloth polishing using 0.5 μm alumina solution. Polished samples were etched with 5% nital solution and microstructures are studied under an optical microscopy (Carl Zeiss made: AXIO Imager A1m) and photographs are taken at 500× magnification. For better clarity, samples were studied under scanning electron microscopy (SEM) (JEOL JSM-5510 with INKA software EDS system using an ultra-thin window detector) and photographs were taken at higher magnifications.

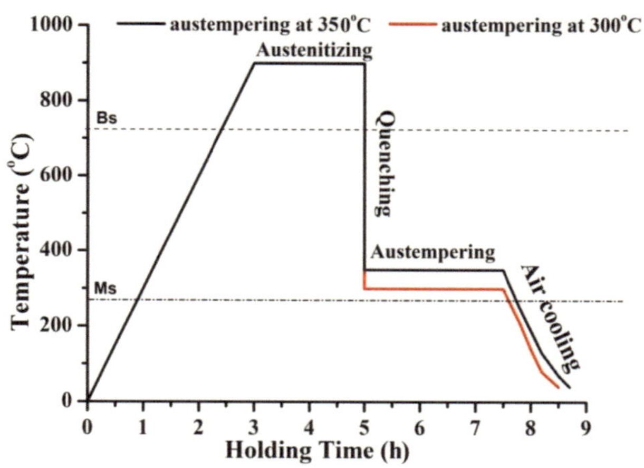

Figure 2.
Typical heat treatment cycle.

X-ray diffraction (XRD) analysis was performed to estimate the volume fraction of retained austenite and its carbon content using anode Co-K_α radiation in 1.79026 targets with 24 kV and tube current was 40 mA. The specified 2θ range was varied from 30 to 110° with a step size of 0.2°/min. Detailed XRD analysis was performed using integrated intensities of the positions and the integrated intensities for the {1 1 1}, {2 2 0} and {3 1 1} planes of FCC austenite as well as the {1 1 0} and {2 1 1} planes of BCC ferrite. The volume fraction of retained austenite was calculated using the following empirical formula [30]:

$$X_\gamma = \frac{I_\gamma / R_\gamma}{(I_\gamma / R\gamma) + (I_\alpha / R_\alpha)} \tag{1}$$

Where I_γ and I_α are the integrated intensities and R_γ and R_α are the theoretical relative intensity for the austenite and ferrite, respectively, and Bainitic ferrite was calculated by using the formula:

$$X_\gamma + X_\alpha + X_g = 1 \tag{2}$$

Where, X_γ and X_α and X_g represent the volume percentage of retained austenite, volume percentage of bainitic ferrite and volume percentage of graphite. The carbon concentration of the austenite was determined using the equation [30].

$$a_\gamma = 0.3548 + 0.0044\,C_\gamma \tag{3}$$

Where a_γ is the lattice parameter of austenite (in nm) and C_γ is the carbon content of austenite (in wt.%). The {1 1 1}, {2 2 0} and {3 1 1} planes of austenite were used to estimate the lattice parameter.

Vickers microhardness test of the weld metals was performed at room temperature using Leco Vickers microhardness tester (Model LM 248SAT) with 100 gf load at 10 s holding. The hardness values were taken from six different positions of each weld specimens and the average of the six values considered the final one.

Tensile properties such as ultimate tensile strength (UTS), yield strength (YS) and % elongation of the welded joints were evaluated using transverse tensile specimen keeping the weld metal at the center of the gauge length. The tests were performed under uniaxial loading at a crosshead speed of 5 mm/min in universal tensile testing m/c (Instron 8862).

Sub-size (55 × 10 × 3.3 mm) and without notch transverse Charpy impact test of the ADI welded joints were performed at room temperature according to ASTM E-23 [31]. Four samples were tested at each austempering condition (300 and 350°C for the 2 h holding time) and an average of four values has been reported.

High cycle fatigue (HCF) test of transverse weld samples as per ASTM E466-15 [32] (**Figure 3**) were performed using Rumul resonant testing machine to determine

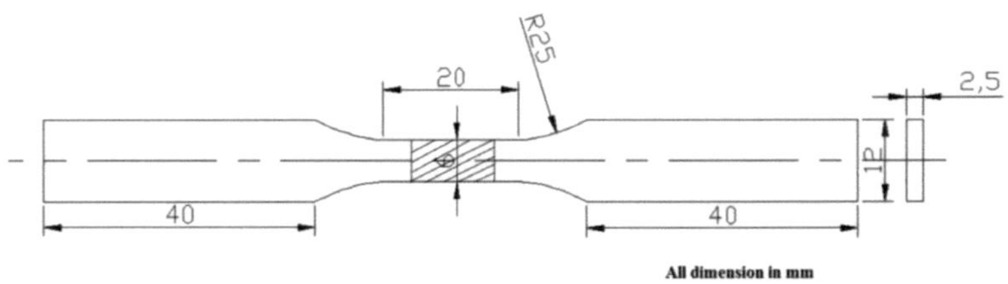

Figure 3.
Schematic view of transverse high cycle fatigue sample as per ASTM 606.

the S-N curve. The tests were run to failure up to 10^7 cycles at constant stress control mode and the number of cycles of failure was recorded with keeping the load ration R at 0.1. The stress levels were varied between 30 and 80% of Yield strength to obtaining the endurance limit and S-N curve was plotted by stress amplitude and the number of cycles in log-log scale.

After successfully testing, fracture surface and crack path of the tested samples were studied under SEM and fractographs were taken at different magnifications.

3. Results and discussion

3.1 Microstructure

3.1.1 Base metal

Figure 4 shows the optical microstructure of as-cast DI (base metal). The microstructure shows graphite nodules surrounded with ferrite matrix. The average nodularity shows 90% with 130 nodules per unit area (mm^2) and average nodule size is r = 18.5 μm.

3.1.2 As-welded microstructure

The optical microstructures of weld metals using two selected coated electrodes containing without and with Ce is shown in **Figure 5**. In **Figure 5a** and **b**, the microstructure shows ledeburitic carbide (LC), alloyed pearlite (AP) and graphite nodules (G). In both the weld metal microstructure shows small amount of graphite nodules with smaller in size due to higher cooling rate experienced in weld metal. Although both the as-weld microstructure shows similar microstructural appearance, a close look into the microstructure reveals difference in grain size and volume percentage of ledeburitic carbide and alloyed pearlite. The presence of Ce in weld metal has caused the structure finer (the finer the dendritic structure), lesser ledeburitic carbide, higher amount of alloyed pearlite and increasing the graphite volume percentage and nodularity.

It has been shown that cerium reduces both primary [33] and secondary [34] dendritic arm spacing as well as inhibit the development of columnar crystal.

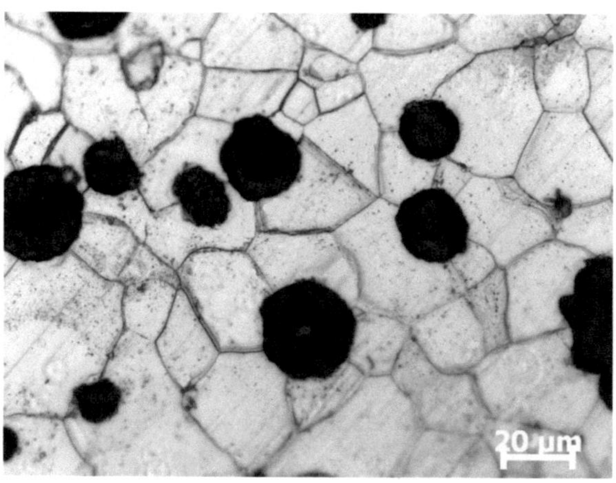

Figure 4.
Optical microstructure of as-cast ductile iron.

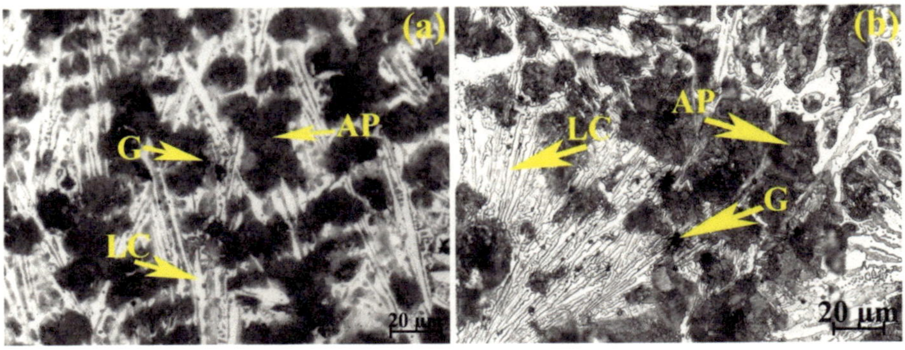

Figure 5.
Optical microstructure of as-welded weld metal (a) Trial 4 (b) Trial 7.

Also, the degree of supercooling for rare earth treated steel has been reported to be smaller than rare earth free steel [35]. The refined microstructure (**Figure 5b**) that has been observed for Ce treated weld metal is presumably due to the fine primary austenite dendrite and suppression of columnar grain growth during solidification of the weld pool. Furthermore, it is believed that smaller degree of super cooling associated with Ce treated weld metal has caused reduction in ledeburitic carbide.

Ce acts as a modifying element on DI as a form of deoxidization and desulfuration [36]. Ce reacts with oxygen and sulfur to form Ce-rich oxides, Ce-rich sulfides or Ce-rich oxide-sulfides formed in DI welds and act as a heterogeneous nuclei of primary carbides, according to the principle of crystallography so that the nuclei of primary carbides can form and grow everywhere in molten metal [37] and refine the structure. Furthermore, cerium content present in the carbide as a form of Ce_2S_3 and CeO_2 (measured by X-RD analysis) and increase the solidification rate to refine the structure [37].

3.1.3 Austempered microstructure

After austempering heat treatment the weld metal microstructure consists of bainitic ferrite (BF) and retained austenite (RA) matrix with graphite nodules (G). **Figure 6a** and **b** illustrate the weld metal microstructure after austempering at 300 and 350°C for 2 h holding time using Trial 4 electrode. Similarly, **Figure 6c** and **d** illustrate weld metal structure after austempering at 300 and 350°C for 2 h holding time using Trial 7 electrode. In both the weld metals austempering at 300°C, the microstructure (**Figure 6a** and **c**) shows needle shape bainitic ferrite, retained austenite and graphite nodules. Whereas at 350°C (**Figure 6b** and **d**) shows feathery shape (lath type) bainitic ferrite with retained austenite and graphite nodules.

For better clarity, the microstructures of heat treated weld metals after austempering at 300 and 350°C for 2 h holding time were studied under SEM and the structures are shown in **Figure** 7 for without and with Ce containing weld metals respectively.

Interestingly, both the weld metal shows the same microstructural appearance at respective austempering conditions. But the structures were varied in morphology, amount, shape and size of bainitic ferrite, amount of retained austenite, nodule size and nodularity with changing the austempering conditions and type of electrode used (without and with Ce containing).

However, at 350°C more amount of retained austenite and lesser amount of bainitic ferrite was observed; but the opposite trend in microstructural constituents has been revealed at 300°C i.e. lower amount of retained austenite and higher amount bainitic ferrite. The nodularity also varied with varying the austempering temperatures and higher nodularity is observed at 350°C at both the weld metals. The microstructural constituents also changed with changing the austempering

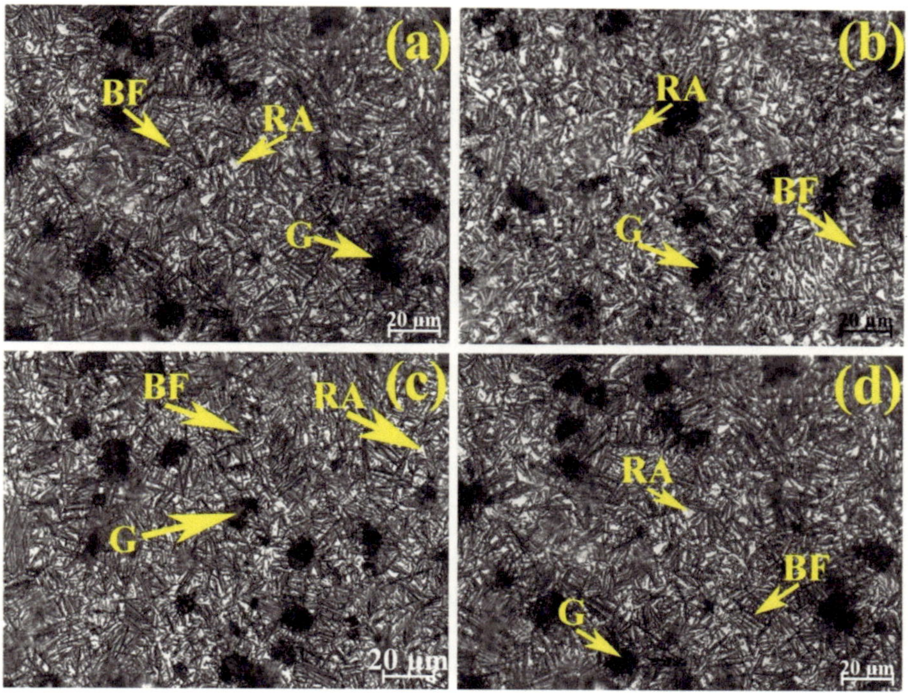

Figure 6.
Optical microstructure of weld metal austempered at (a) 300°C (b) 350°C for 2 h holding time using Trial 4 and (c) 300°C (d) 350°C for 2 h holding time using Trial 7 coated electrode.

holding time at a given temperature [9]. However, for both the austempering temperatures the variation of the microstructural constituent is similar i.e. with changing the holding time from 1.5 to 2 h the amount of retained austenite was increased and the amount of bainitic ferrite was decreased also refine the bainitic

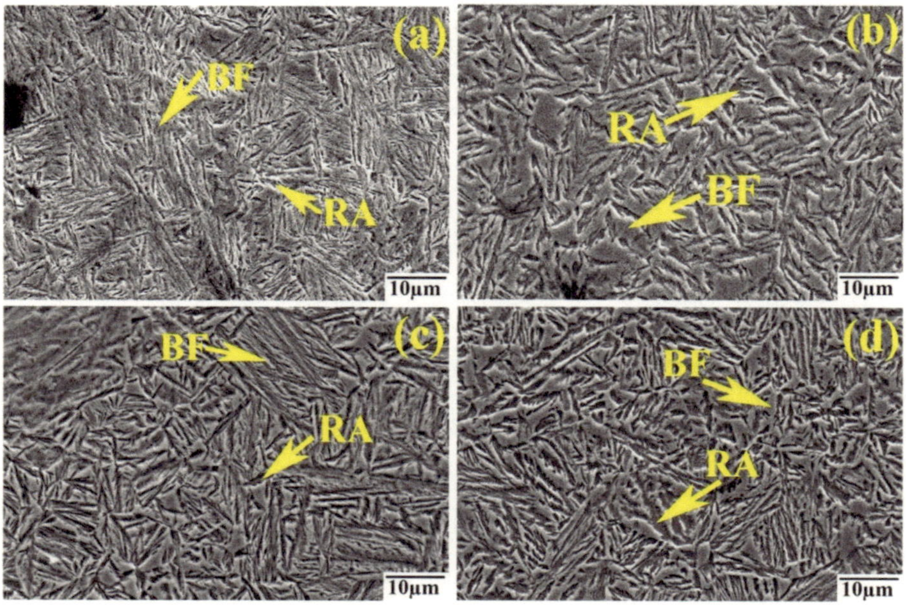

Figure 7.
SEM microstructure of weld metal austempered at (a) 300°C (b) 350°C for 2 h holding time using Trial 4 and (c) 300°C (d) 350°C for 2 h holding time using Trial 7 coated electrodes.

ferrite shape and size. Further increasing the austempering holding time from 2 to 2.5 h the amount of retained austenite was decreased and the amount of bainitic ferrite was increased. Interestingly, at both 300 and 350°C austempering temperature for 2 h holding time the carbon enrichment in austenite is maximum (**Figure 9**) which has caused to stabilize more amount of retained austenite (**Figure 8**) at both the austempering conditions after cooling to room temperature. However, at higher holding time (2.5 h) untransformed austenite transformed to carbides (ε carbide) and bainitic ferrite leading to decrease the amount of retained austenite content [1].

The observed finer and homogeneous structure along with increasing the amount of retained austenite (**Figure 8**) with the presence of Ce content in weld metal. At both 300 and 350°C austempering temperature microstructure attributed finer the bainitic ferrite size, higher amount of carbon enriched retained austenite and higher graphite nodularity was observed with presence of Ce in weld metal. The carbon enrichment of austenite will be faster in Ce treated weld metal due to lesser diffusion distance for carbon, which diffuses from fine cementite lamellae of pearlite. Also, smaller the nodule size having more surface area to volume ratio will favor carbon diffusion from graphite [38]. Thus, with the increase of carbon content of initial austenite the driving force of stage I reaction become slow and delay the transformation rate of bainitic ferrite due to drag effects of Ce. As a result more amount of carbon was diffused to the surrounding austenite and austenite become more stable.

3.1.4 Volume percentage of retained austenite and its carbon content

The volume percentage of retained austenite and its carbon content of both the weld metals after austempering at 300 and 350°C for different holding times have been calculated from X-RD analysis. The variation of retained austenite with changing the holding time (1.5, 2 and 2.5 h) at 300 and 350°C austempering temperature has been plotted in **Figure 8**. In **Figure 8**, it is seen that at both 300 and 350°C temperature both the weld metal (without and with Ce containing) the amount of retained austenite was changed with changing the austempering holding time. Holding time changed from 1.5 to 2 h the amount of retained austenite was increased with decreasing the amount of bainitic ferrite. With further increases the holding time from 2 to 2.5 h the amount of retained austenite was decreased. Although the nature of change the amount of retained austenite is same at both

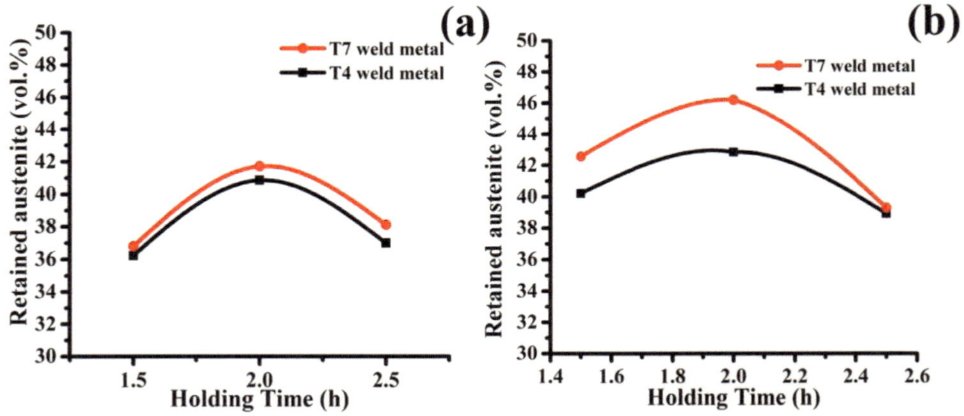

Figure 8.
Volume percentage of retained austenite content at different holding time of weld metal austempering at (a) 300°C and (b) 350°C using Trial 4 and Trial 7 electrodes.

the austempering temperature of both the weld metals, austempering at 350°C shows the higher amount of retained austenite compared to 300°C at each holding time. This could be due to the higher diffusion rate of carbon during bainitic ferrite transformation at higher austempering temperature [39].

Bainitic ferrite formation is a growth process; during austempering process bainitic ferrite is transformed from the existing austenite (residual austenite). During the transformation of bainitic ferrite, carbon was diffused from the bainitic ferrite to the surrounding austenite to make the austenite stable, and this austenite is called untransformed austenite or retained austenite. At lower austempering temperature, due to higher super cooling the transformation rate of bainitic ferrite is high and diffused less amount of carbon to the surrounding austenite, as a result formed less amount of retained austenite in weld metals. Furthermore, at higher austempering temperatures, due to lower supper cooling the growth of bainitic ferrite is high and diffused higher amount of carbon content to the surrounding austenite and shows higher amount of retained austenite in weld metals.

Ce is a modifying alloying element and act as a nodularizing and austempering element during austempering process. The presence of Ce in weld metal to slow the austempering kinetics and prolongs to the stage I reaction. As a result higher amount of carbon was diffused during the transformation of bainitic ferrite to the surrounding austenite and show a higher amount of retained austenite at both the austempering temperature than without Ce content weld metal also refine the microstructure to indicate higher toughness and longer fatigue life.

Furthermore, the carbon content of retained austenite in two weld metals (without and with Ce containing) at 300 and 350°C for 1.5, 2 and 2.5 h holding time have been determined using the empirical formula (2). The calculated results of the amount of carbon present in austenite at 300 and 350°C for different holding times have been plotted in **Figure 9**. In **Figure 9**, it is seen that, in both the weld metals, the trend in variation of carbon content in retained austenite with changing the austempering holding at 300 and 350°C is similar to the variation of retained austenite (**Figure 8**). The maximum amount of carbon content was achieved at 2 h holding time irrespective of the austempering temperature and electrode composition. Weld metal containing Ce show the highest carbon content (2.2 wt.%) in retained austenite at 350°C for 2 h holding time.

During austempering transformation, bainitic ferrites are nucleated out of austenite (residual austenite) to refusing the carbon content to the surrounding austenite

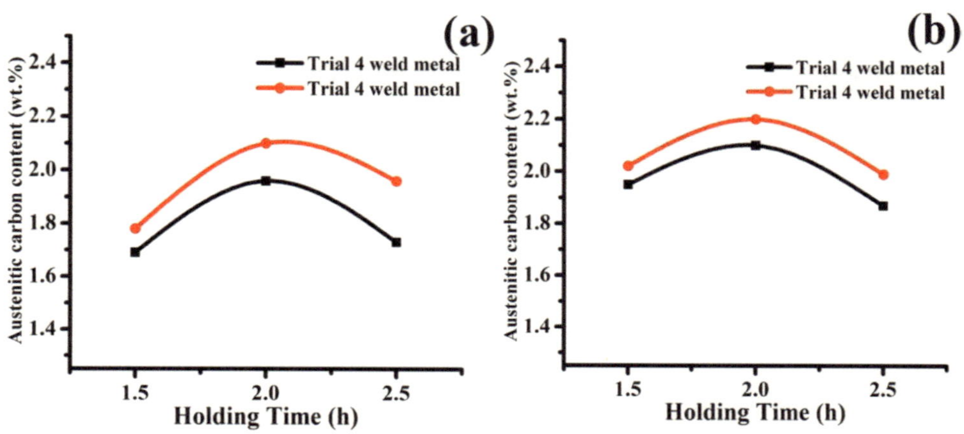

Figure 9.
Austenitic carbon content of weld metal at different holding time austempering at (a) 300°C and (b) 350°C using Trial 4 and Trial 7 electrodes.

and austenite become stabilized. Thus, maximum stability of retained austenite should possess at 2 h holding time, irrespective of austempering temperature of both the weld metals.

3.2 Mechanical properties

The mechanical properties of ADI welded joints depends on the amount of retained austenite and its carbon content presence in microstructure [1, 2, 40]. Mechanical properties were varied with varying the austempering temperature and it's holding time due to the variation of microstructural constitutes, shape and size, nodularity and numbers of nodule presence in per mm^2. At 300°C austempering temperature microstructure consists of needle shaped bainitic ferrite with small amount of retained austenite indicates higher tensile strength and hardness with lower the toughness. However, at 350°C austempering temperature structure shows feathery shaped bainitic ferrite with higher amount of retained austenite indicated lower tensile strength and hardness with higher toughness and fatigue strength. The average mechanical properties of both the weld metals (Trial 4 and Trial 7) after austempering at 300 and 350°C for 2 h holding time are given in **Table 4**.

3.2.1 Microhardness

The microhardness of both the weld metals (without and with Ce containing) after austempering at 300 and 350°C for 1.5, 2 and 2.5 h holding time were taken at six different positions of the weld metal and the average of six considered final value. The average microhardness of the weld metals at both 300 and 350°C austempering temperature are plotted in **Figure 10** with respect to the holding time. **Figure 10** demonstrates that at 300°C both the weld metal shows higher hardness value than 350°C due to the variation of the amount of carbon enriched austenite. However, the hardness values are changed with changing the austempering holding time. In both the weld metals at both 300 and 350°C holding time changed from 1.5 to 2 h, the hardness values are decreased with increasing the amount of retained austenite. With increasing holding time from 2 to 2.5 h the hardness values are increased, with decreasing the amount of carbon enriched retained austenite in weld metals. At higher holding time the stage II reaction was started and the carbon enriched retained was diffused to bainitic ferrite and carbide, as a results the amount of retained austenite was decreased and the hardness value increased [41]. At both the austempering temperature of both weld metals at 2 h holding time attributed lower hardness value, the presence of higher amount of retained austenite.

Further, at each austempering conditions Ce containing weld metal shows lower hardness value due to higher amount of carbon enriched retained austenite in weld

Electrode	Austempering temperature (°C)	Holding time (h)	Charpy impact (J)	Fatigue strength (MPa)	Microhardness (Hv)
Trial 4 (without Ce)	300	2	20	211	327
	350		26	241	280
Trial 7 (with Ce)	300		22	241	290
	350		31	302	273

Table 4.
Mechanical properties of ADI weld metals.

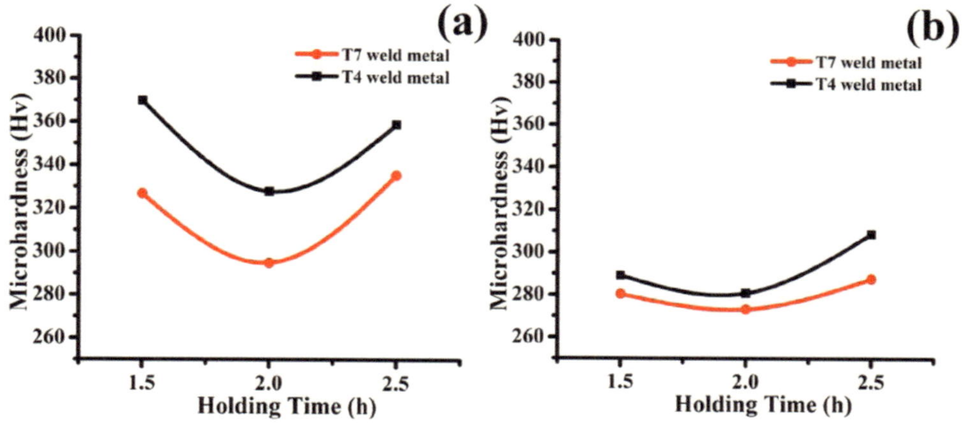

Figure 10.
Variation of microhardness of weld metals at different holding time austempering at (a) 300°C and (b) 350°C using Trial 4 and Trial 7 electrodes.

metal and attributed minimum hardness (273Hv) at 350°C for 2 h holding time presence of higher amount of retained austenite (46.7%) compared to without Ce content weld metal.

3.2.2 Tensile test

The transverse tensile tests of Trial 4 and Trial 7 weld joints after austempering at 300 and 350°C for 2 h holding time were carried out. For both the austempering conditions of both the weld metals failure of the test samples took place from the base metal as shown in **Figure 11**, indicating that weld metal is stronger than base metal. Thus the transverse tensile results show 100% joint efficiency of ADI weld metals (at 300 and 350°C for 2 h holding time) indicating successfully development of coated electrode and establishment of a weld procedure which could be applied commercially.

The tensile properties of ADI is strongly depends on the amount of retained austenite and its carbon content, the shape and size of bainitic ferrite and the numbers of graphite nodules presence in per unit area [42, 43]. At each austempering conditions both the weld metal attributed more amount of retained austenite and finer bainitic ferrite and smaller size of graphite nodules compare to base metal and HAZ. During tensile testing weld metal is possibly more strain hardened compare to base metal and HAZ due to presence of higher amount of retained austenite [43, 44]. It is well known that austenite having FCC structure possesses higher strain hardening rate than BCC ferrite (bainitic ferrite) and the strain hardening rate of austenite also increases with presence of carbon content [45]. Thus, one would expect, due to more amount of retained austenite along with higher carbon, that tensile strength of Ce containing weld metal would be higher than weld metal having without Ce at 350°C austempering temperature for 2 h holding time.

3.2.3 Charpy impact test

The average sub-size Charpy impact values of without and with Ce containing weld metals after austempering at 300 and 350°C for 2 h holding time are given in **Table 4**. The result states that the Charpy impact values are increased with increasing the austempering temperature from 300 to 350°C. The Charpy impact values are strongly depends on the amount of retained austenite, shape and size of bainitic

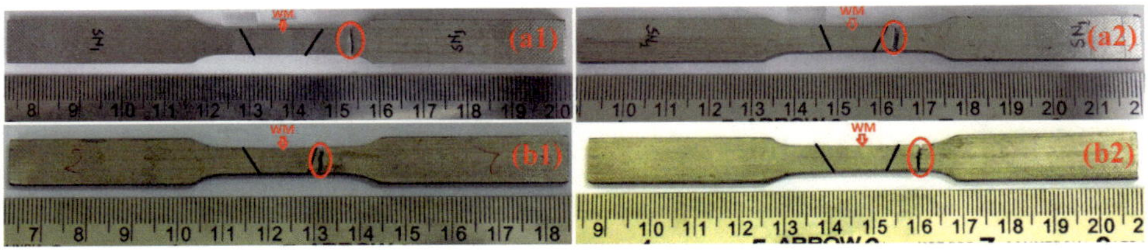

Figure 11.
Transverse tensile samples after testing, austempering at (a1) 300°C, (a2) 350°C for 2 h using Trial 4 and (b1) 300°C, (b2) 350°C for 2 h using trial electrode.

ferrite, graphite nodularity, size of the nodules and numbers of nodules presence in per unit area. Presence of higher carbon content in austenite the strain hardening rate of austenite is high and consequently more energy is being absorbed leading to increase in impact toughness [42, 43]. The higher nodularity could suppress the crack initiation due to lower stress concentration and increase in amount of graphite nodules could act as crack arrester during impact testing. Therefore, increasing the graphite nodularity of ADI can improved the high cycle fatigue strength. In both the weld metals at 350°C to attributed higher carbon enriched retained austenite to signify the higher impact values.

Furthermore Ce containing weld metal shows higher Charpy impact values at both the austempering temperature. Ce content in weld metal to refine the bainitic ferrite and retained austenite shape with increased the amount of retained austenite content, smaller the graphite nodules with higher nodularity indicates higher impact values. The highest impact value (31 J) was obtained at 350°C at Ce content weld metal presence of maximum amount (46%) of carbon enriched retained austenite.

The fracture surfaces of the freshly broken Charpy impact test specimens of both Trial 4 and Trial 7 after austempering at 300 and 350°C for 2 h holding time were examined under SEM in order to relate impact properties to operative fracture mechanism and are given in **Figure 12**. At 300°C both the weld metal (**Figure 12a** and **c**) fracture surface exhibits predominantly dimple and quasi-cleavage types fracture. However, at 350°C (**Figure 12b** and **d**) the fracture surface exhibits predominantly

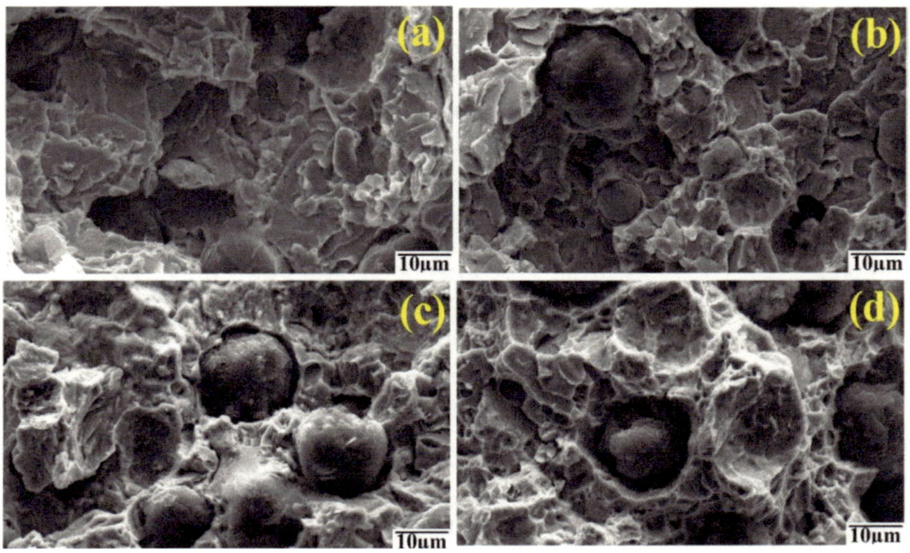

Figure 12.
Fracture surface of ADI weld metal austempering at (a) 300°C and (b) 350°C for 2 h using Trial 4 and (c) 300°C and (d) 350°C for 2 h using Trial 7 developed electrode.

dimple in nature ductile fracture of both the weld metals. The presence of Ce content in weld metal the sizes of the dimple are more fine and with changing the austempering temperature (300–350°C) the shape and size of dimple are more finer to indicate the higher impact value. It appears that extensive plastic flow around the graphite nodules results in stress concentration at graphite-matrix interface which leads to a decohesion between graphite nodules and the matrix. **Figure 12** illustrates the small graphite nodules within cavities or dimples that initiated the microvoids. Subsequent growth and coalescence of these microvoids produced dimple rupture network. The crack, which is likely to initiate near the graphite nodule, propagates through the matrix to reach the adjoining nodules. It is anticipated that plastic deformation in the matrix ahead of the regions of decohesion will thus be confined essentially to the soft ferrite regions [42] and the crack propagation becomes difficult across the tough austenite to join up with similar micro cracks in the neighboring ferrite needles. Therefore, one can expect improved toughness with increasing the volume fraction of retained austenite.

3.2.4 High cycle fatigue analysis

The fatigue life of weld joints after austempering at 300 and 350°C using without and with Ce containing coated electrodes are represented through S-N curve in **Figure 13**. The S-N curve is drawn through the data point as best fit line and the arrows are indicates the samples did not fail before an excess of 10^7 cycle or more stress cycle to called the endurance limit. The maximum strength obtained at 10^7 cycles is called the fatigue strength. In **Figure 13** illustrate that the fatigue strength of the weld joints were varied with varying the austempering temperature and respective weld metal composition. The maximum fatigue strength (302 MPa) was obtained at 350°C for with Ce weld metals due to higher toughness and lower tensile strength and hardness of weld metal (**Table 4**).

At lower (300°C) austempering temperature microstructure shows needle shape bainitic ferrite with lower amount of retained austenite to indicate higher tensile strength and hardness with lower impact toughness. However, at higher (350°C) austempering temperature microstructure shows coarser bainitic ferrite with higher amount of retained austenite to indicates lower tensile value and hardness with higher impact toughness indicates higher fatigue strength.

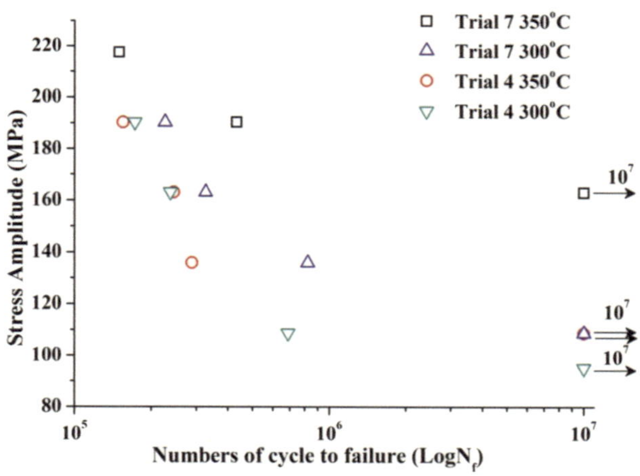

Figure 13.
Comparison of S-N curve of high cycle fatigue test for Trial 4 and Trial 7 weld metals at different austempering conditions (arrow indicating the endurance limit of the each condition).

The fatigue strength of ADI joints depended on the amount of retained austenite and its carbon content presence in ADI weld metals and the values are increased with increasing the amount of retained austenite content due to the higher strain hardening behavior of austenite [46]. Owing to the higher strain hardening behavior of austenite to delay the formation of persistent slip bands and reduces the nucleating of fatigue crack growth. However higher amount of retained austenite transformed to plastic induced martensite during high cycle fatigue test [26, 46, 47]. Such types of transformation was occurred at the plastic zones to ahead the fatigue crack and relax the stress concentration at the crack tip, as a result to reduce the fatigue crack growth rate and increase the fracture toughness.

However, the fatigue strength also depends on the shape and size of the bainitic ferrite, retained austenite and graphite nodules. The fatigue strength of ADI was improved when the nodule count was increased, in particular at austempered temperatures for higher toughness [26]. Therefore, ADI with a higher nodule count with the smaller size exhibits a better fatigue life [26]. The more pronounced graphite nodule size with higher optimum toughness to exhibits higher amount of retained austenite indicates maximum fatigue strength.

Presence of Ce content in weld metal illustrates higher fatigue strength than without Ce content weld metal. Ce content to refine the austenitic grain size also increasing the amount of retained austenite and its carbon content to illustrate the higher fatigue life. The smaller retained austenite grain size apparently creates more dislocation barrier; as a result the formation of precipitate slip band is more difficult and required more number of cycles to initiate the fatigue crack [14]. The maximum fatigue strength was obtained at 350°C at Ce containing weld metal, presence of maximum amount of retained austenite content (46.7%).

The fracture surface of welded fatigue test specimen after austempering at 300 and 350°C for 2 h holding time are analyzed by SEM studies and the fractographs are shown in **Figure 14**. In **Figure 14**, the fracture surface reveals ductile fracture with dimple formation also some cleavage fracture which is mostly

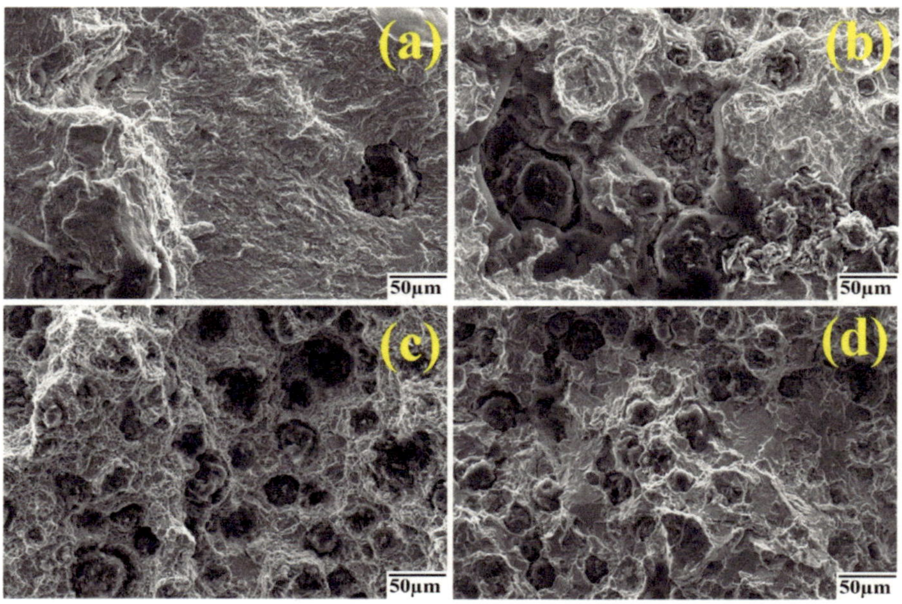

Figure 14.
Fatigue fracture surface austempered ADI weld metals at (a) lower strength and (b) higher strength using Trial 4 electrode and (c) lower strength and (b) higher strength using Trial 7 electrode.

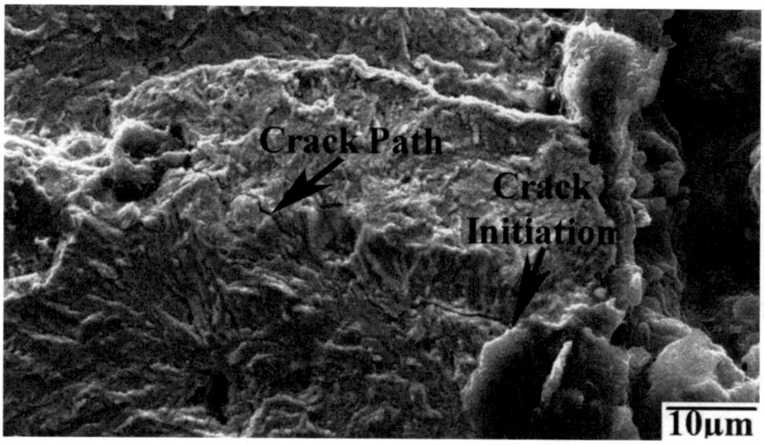

Figure 15.
Fatigue fracture at ADI weld metals indicating the crack initiation and crack propagation path.

occurs in the vicinity of graphite nodules. The combination of ductile striation and cleavage plane whose river patterns go in to tear rivers is named as quasi cleavage [47]. At 300°C (**Figure 14a** and **c**) austempering temperature shoes mixture of dimple and quasi-cleavage fracture, however at 350°C shows (**Figure 14b** and **d**) fully dimple in nature feature is the dominant fracture mechanism to indicate ductile or transgranular in nature. Presence of Ce to refined the fracture surface and shows smaller size dimple in fracture surface and improved the fatigue strength.

The fatigue crack (**Figure 15**) was initiated from the interface between the matrix and graphite nodules and crack path through the least resistance matrix phase due to the week interface bond between graphite and matrix also the lower elastic modulus of graphite. The crack path preferentially intersects with graphite nodules and there is an apparent crack branching mechanism to relate the crack-nodule interactions. This mechanism is responsible for decreasing the fracture energy and reduce the crack propagation rate [48]. **Figure 15** shows, the fatigue crack was initiated comparatively larger size graphite nodules and propagates to the nearest graphite through the matrix surface. The graphite nodules are not perfectly spheroid and the interface between the matrix and graphite are irregulars with multiple sharp corner and high stress concentration to constitute crack that emanate from the graphite nodules [48]. However, the fatigue crack propagation path depends on the next graphite nodules ahead of the crack tip [48] and the crack front connect to the graphite nodules along to the crack path. At higher austempering temperature (350°C), decohesion of nodules and microcracks was observed around the graphite nodules.

It is important to note that several graphite nodules are involved with the growth of crack front, the shape size and distribution of the graphite nodules affects the average growth of crack [49].

Previous investigators [17, 26, 50, 51] states the crack was initiated at the interface between the graphite nodules, and a matrix. However crack also may initiate defects presence in metal such as inclusion, shrinkage and irregular shaped of graphite nodules [52, 53]. The nucleation of the crack grows very rapidly after initiation of crack and decelerates due to the interaction of the matrix structure. The crack is strongly influenced by the microstructure in the crack path region and propagated through discontinuous manner due to the retardation of the matrix grain boundary.

4. Conclusion

Based on the above studies, the following conclusions can be drawn.

1. Both Trial 4 (without Ce) and Trial 7 (with Ce) weld metals are responds in austempering heat treatment at 300 and 350°C for 1.5, 2 and 2.5 h holding time. At 300°C microstructure shows needle shape bainitic ferrite with retained austenite however at 350°C shows feathery shaped bainitic ferrite with retained austenite. Ce content in weld metal to refine the bainitic ferrite and austenite grain size and structures become homogeneous.

2. Volume percentage of retained austenite and its carbon content changed with changing the austempering temperature and holding time. At 300°C shows less amount of retained austenite with higher amount of bainitic ferrite. However at 350°C shows higher amount of retained austenite with comparatively less amount of bainitic ferrite. Retained austenite also varied with varying the austempering holding time at respective temperature and shows maximum retained austenite content at 2 h holding time at each austempering temperature for both the weld metals. However, Ce content in weld metal to enlarge the process windows to delay the stage I reaction and shows more amount of retained austenite at each austempering temperature.

3. The microhardness of weld metal shows lower value at 350°C, presence of higher amount of retained austenite than 300°C austempering temperature. With changing the austempering holding time, the hardness values are changed and show lower hardness value at 2 h holding time with respective austempering temperature. Ce content in weld metal decreased the hardness value and shows at both the temperature, presence of higher amount retained austenite and homogeneous microstructure. Tensile test result shows failure take place from the base metal for both Trial 4 and Trial 7 weld metals at 300 and 350°C for 2 h holding time to indicate the 100% joint efficiency.

4. Charpy impact values are changes with changing the austempering temperature with changing the amount of retained austenite in weld metal. 350°C shows higher impact value for both weld metals presence of higher amount of retained austenite. Ce in weld metal to improve the Charpy impact value as a result of refine the microstructure and increasing the amount of retained austenite. Maximum Charpy impact value shows at Trial 7 weld metal at 350°C austempering conditions.

5. Fatigue strength of both welded joints was improved with improving the austempering temperature. At 350°C feathery shaped bainitic ferrite with higher amount of retained austenite illustrate higher fatigue strength than needle shaped bainitic ferrite with small amount of retained austenite at 300°C. Presence of Ce content in weld metal Trial 7 weld metal shows higher fatigue strength at both austempering temperature due to lower hardness and higher Charpy impact value.

6. Fatigue fracture surface shows at 300°C temperature predominantly mixture with dimple and reverse pattern quasi-cleavage types fracture, however at 350°C shows dimple types fracture to indicates ductile in nature. Ce content

in weld metal to refine the dimple size indicates higher toughness. Crack was initiated in comparatively larger size graphite nodules and propagated through less resistance matrix to the nearest graphite.

Author details

Tapan Sarkar
Metallurgical and Material Engineering Department, Welding Technology Centre, Jadavpur University, Kolkata, India

*Address all correspondence to: tapansarkar.met@gmail.com

References

[1] Panneerselvam S, Putatunda SK, Gundlach R, Boileau J. Influence of intercritical austempering on the microstructure and mechanical properties of austempered ductile cast iron (ADI). Materials Science and Engineering A. 2017;**694**:72-80

[2] Acharya PP, Udupa R, Bhat R. Microstructure and mechanical properties of austempered AISI 9255 high-silicon steel. Materials Science and Technology. 2018;**34**(3):355-365

[3] Bosnjak B, Radulovic B, Pop-Toner K, Asanovic V. Influence of microalloying and heat treatment on the kinetics of bainitic reaction in austempered ductile iron. Journal of Materials Engineering and Performance. 2001;**10**(2):203-211

[4] Lin SC, Lui TS, Chen LH, Song JM. Effect of pearlite on the vibration-fracture behavior of spheroidal graphite cast irons under resonant conditions. Metallurgical and Materials Transactions A. 2002;**33**(8):2623-2634

[5] Lin CK, Lee WJ. Effects of highly stressed volume on fatigue strength of austempered ductile irons. International Journal of Fatigue. 1998;**20**(4):301-307

[6] Chapetti MD. High-cycle fatigue of austempered ductile iron (ADI). International Journal of Fatigue. 2007;**29**(5):860-868

[7] Gundlach RB, Janowak JF. Austempered ductile iron combines strength with toughness and ductility. Metal Progress. 1985;**128**(2):19-26

[8] Ayman H, Elsayed MM, Megahed AA, Sadek KM. Fracture toughness characterization of austempered ductile iron produced using both conventional and two-step austempering processes. Materials and Design. 2009;**30**(6):1866-1877

[9] Panneerselvam S, Martis CJ, Putatunda SK, Boileau JM. Fracture toughness characterization of austempered ductile iron produced using both conventional and two-step austempering processes. Materials Science and Engineering A. 2015;**626**(6):237-246

[10] Putatunda SK. Development of austempered ductile cast iron (ADI) with simultaneous high yield strength and fracture toughness by a novel two-step austempering process. Materials Science and Engineering A. 2001;**315**(1-2):70-80

[11] Kim YJ, Shin H, Park H, Lim JD. Investigation into mechanical properties of austempered ductile cast iron (ADI) in accordance with austempering temperature. Materials Letters. 2008;**62**(3):357-360

[12] Grech M, Young JM. Influence of austempering temperature on the characteristics of austempered ductile iron alloyed with copper and nickel. Transactions of the American Foundrymen's Society. 1990;**98**:345-352

[13] Bartosiewicz L, Krause AR, Alberts FA, Singh I, Putatunda SK. Influence of microstructure on high-cycle fatigue behavior of austempered ductile cast iron. Materials Characterization. 1993;**30**(4):221-234

[14] Puspitasari P, Gapsari F. Fatigue crack growth behavior of nodular cast iron subjected to two-step austempering. Metallurgia. 2018;**57**(4):317-320

[15] Zhang J, Song Q, Zhang N, Lu L, Zhang M, Cui G. Very high cycle fatigue property of high-strength austempered ductile iron at conventional and ultrasonic frequency loading. International Journal of Fatigue. 2015;**70**:235-240

[16] Bahmani M, Elliot R, Varahram N. The relationship between fatigue strength and microstructure in an austempered Cu-Ni-Mn-Mo alloyed ductile iron. Journal of Materials Science. 1997;**32**(20):5383-5388

[17] Shirani M, Härkegård G. Fatigue life distribution and size effect in ductile cast iron for wind turbine components. Engineering Failure Analysis. 2011;**18**(2):1496-1510

[18] Sofue M, Okada S, Sasaki T. High-quality ductile cast iron with improved fatigue strength. In: 82nd Annual Meeting. 1978. pp. 173-182

[19] Diao X, Ning Z, Cao F, Sun J. Effect of CE on graphite nodule count and size distribution in ductile iron. International Journal of Modern Physics B. 2009;**23**(6-7):1853-1860

[20] Choi JO, Kim JY, Choi CO, Kim JK, Rohatgi PK. Effect of rare earth element on microstructure formation and mechanical properties of thin wall ductile iron castings. Materials Science and Engineering A. 2004;**383**(2):323-333

[21] El-Banna EM. Effect of preheat on welding of ductile cast iron. Materials Letters. 1999;**41**(1):20-26

[22] Imasogie BI, Wendt U. Characterization of graphite particle shape in spheroidal graphite iron using a computer-based image analyzer. Journal of Minerals and Materials Characterization and Engineering. 2004;**3**(1):1-12

[23] Sarkar T, Pal TK. Response of austempering heat treatment on microstructure and mechanical property in different zones of as-welded ductile iron (DI). SAE International Journal of Materials and Manufacturing. 2018;**11**(2):151-159

[24] Sarkar T, Pal TK, Pramanick AK. Variation of nano Ce content in austempered ductile iron (adi) weld metal and its effects on microstructure. Transactions of the Indian Institute of Metals. 2018;**71**(10):2375-2385

[25] Zhang J, Li W, Song Q, Zhang N, Lu L. Fatigue of austempered ductile iron with two strength grades in very high cycle regime. Journal of Materials Engineering and Performance. 2016;**25**(3):744-749

[26] Lin CK, Lai PK, Shih TS. Influence of microstructure on the fatigue properties of austempered ductile irons-I. High-cycle fatigue. International Journal of Fatigue. 1996;**18**(5):297-307

[27] Sarkar T, Pramanick AK, Pal TK. Some aspects on the welding characteristics and formation of microstructures in a newly developed coated electrode for austempered ductile iron (ADI). Indian Welding Journal. 2015;**48**(4):44-60

[28] Sarkar T, Pramanick AK, Pal TK, Pramanick AK. Development of a new coated electrode with low nickel content for welding ductile iron and its response to austempering. International Journal of Minerals, Metallurgy, and Materials. 2018;**25**(9):1090-1103

[29] Guide for Welding Iron Castings, ANSI/AWS D11.2-89 (R2006). An American National Standard Institute; 1988. pp. 52-54

[30] Cullity BD. Elements of X-Ray Diffraction. Reading USA:Addison and Wesley Publishing Company Inc; 1978. pp. 32-106

[31] ASTM E23-16b. Standard Test Methods for Notched Bar Impact Testing of Metallic Materials. West Conshohocken, PA: ASTM International; 2016

[32] ASTM E466-15. Standard Practice for Conducting Force Controlled Constant Amplitude Axial Fatigue Tests of Metallic Materials.

West Conshohocken, PA: ASTM International; 2015

[33] Nuri Y, Tetsuro O, Takeshi H, Osamu K. Solidification microstructure of ingots and continuously cast slabs treated with rare earth metal. Transactions of the Iron and Steel Institute of Japan. 1982;**22**:399-407

[34] Xu G, Xiao J. New frontiers in rare earth science and applications. In: Proceedings of the International Conference on Rare Earth Development and Applications. Beijing the People's Republic of China Academic Press; 10-14 September 1985; 1985

[35] Jiang MZ, Yu YC, Li H, Ren X, Wang SB. Effect of rare earth cerium addition on microstructures and mechanical properties of low carbon high manganese steels. High Temperature Materials and Processes. 2017;**36**(2):145-153

[36] Qingxiang Y, Bo L, Jianhua L, Mei Y. Effect of rare earth elements on carbide morphology and phase transformation dynamics of high Ni-Cr alloy cast iron. Journal of Rare Earths. 1998;**16**:36-40

[37] Zhi X, Liu J, Xing J, Shengqiang M. Effect of cerium modification on microstructure and properties of hypereutectic high chromium cast iron. Materials Science and Engineering A. 2014;**603**:98-103

[38] Fernandino DO, Massone JM, Boeri RE. Characterization of the austemperability of partially austenitized ductile iron. Journal of Materials Processing Technology. 2013;**213**(10):1801-1809

[39] Mallia J, Grech M. Effect of silicon content on impact properties of austempered ductile iron. Materials Science and Technology. 1997;**13**(5):408-414

[40] Abioye AA, Atanda PO, Abioye OP, Afolalu SA, Dirisu JF. Microstructural characterization and some mechanical behaviour of low manganese austempered ferritic ductile iron. International Journal of Applied Engineering Research. 2017;**12**(23):13435-13441

[41] Putatunda SK, Jiang HY. A novel processing of austempered ductile cast iron (ADI). Materials Science Forum. 2003;**426**:913-918

[42] Rao PP, Putatunda SK. Dependence of fracture toughness of austempered ductile iron on austempering temperature. Metallurgical and Materials Transactions A. 1998;**29**(12):3005-3016

[43] Wang B, Barber G, Sun X, Shaw M, Seaton P. Characteristics of the transformation of retained austenite in tempered austempered ductile iron. Journal of Materials Engineering and Performance. 2017;**26**(5):2095-2101

[44] Putatunda SK, Martis C. Influence of two-step austempering process on the fracture toughness of a low carbon low alloy (LCLA) steel. Materials Science Forum. 2012;**706**:2259-2264

[45] Batra U, Ray S, Prabhakar SR. Impact properties of copper-alloyed and nickel-copper alloyed ADI. Journal of Materials Engineering and Performance. 2007;**16**(4):485-489

[46] Li XH, Saal P, Gan WM, Hoelzel M, Volk W, Petry W, et al. Strain-induced martensitic transformation kinetic in austempered ductile iron (ADI). Metallurgical and Materials Transactions A. 2018;**49**(1):94-104

[47] Lin CK, Wei JY. High-cycle fatigue of austempered ductile irons in various-sized Y-block castings. Materials Transactions, JIM. 1997;**38**(8):682-691

[48] Bullooch JH. Near threshold fatigue behaviour of flake graphite cast irons microstructures. Theoretical

and Applied Fracture Mechanics. 1995;**24**(1):65-78

[49] Greno GL, Otegui JL, Boeri RE. Mechanisms of fatigue crack growth in austempered ductile iron. International Journal of Fatigue. 1999;**21**(1):35-43

[50] Stokes B, Gao N, Reed PAS. Effects of graphite nodules on crack growth behaviour of austempered ductile iron. Materials Science and Engineering A. 2007;**445**:374-385

[51] Jen KP, Wu J, Kim S. The study of fracture and fatigue behavior of austempered ductile iron. AFS Trans. 1992;**92**:133-164

[52] Lin CK, Fu CS. Low-cycle fatigue of austempered ductile irons in various-sized Y-block castings. Materials Transactions, JIM. 1997;**38**(8):692-700

[53] Lin CK, Chang CW. Influence of heat treatment on fatigue crack growth of austempered ductile iron. Journal of Materials Science. 2002;**37**(4):709-716

Chapter 9

Spark Plasma Sintered High-Entropy Alloys: An Advanced Material for Aerospace Applications

Ayodeji Ebenezer Afolabi, Abimbola Patricia I. Popoola and Olawale M. Popoola

Abstract

High-entropy alloys (HEAs) are materials of high property profiles with enhanced strength-to-weight ratios and high temperature-stress-fatigue capability as well as strong oxidation resistance strength. HEAs are multi-powder-based materials whose microstructural and mechanical properties rely strongly on stoichiometry combination of powders as well as the consolidation techniques. Spark plasma sintering (SPS) has a notable processing edge in processing HEAs due to its fast heating schedule at relatively lower temperature and short sintering time. Therefore, major challenges such as grain growth, porosity, and cracking normally encountered in conventional consolidation like casting are bypassed to produce HEAs with good densification. SPS parameters such as heating rate, temperature, pressure, and holding time can be utilized as design criteria in software like Minitab during design of experiment (DOE) to select a wide range of values at which the HEAs may be produced as well as to model the output data collected from mechanical characterization. In addition to this, the temperature-stress-fatigue response of developed HEAs can be analyzed using finite element analysis (FEA) to have an in-depth understanding of the detail of inter-atomic interactions that inform the inherent material properties.

Keywords: spark plasma sintering, high entropy alloys, advance material, aerospace application

1. Introduction

1.1 Background/motivation

In the aerospace industry, the landing gear is essential for support during landing and ground operations in an aircraft. Manufacturers attach the landing gear to primary structural members of an aircraft and must bear the heavy compressive load. Therefore, when landing, the impact energy of the aircraft will be absorbed through the design and a minimized load transmitted to the airframe. Thus, materials used for landing gear application must be able to withstand high operating temperatures, fatigue, creep, cyclic and translational movement of parts at high-speed,

IntechOpen

and chemical, erosion, wear, and oxidation degradations [1–3]. The development of high-performance materials with superior characteristics for aerospace application has continually been the challenge faced by material engineers and scientists over the years. However, materials engineers constantly create and improve properties of materials by applying the existing knowledge of science to develop advanced engineering materials exhibiting better service performance through the development of various manufacturing techniques for different applications as shown in **Figure 1** [4–6]. The traditional alloys used for commercial purposes were designed by choosing a core element which made up the matrix of the part and addition of elemental solutes to the primary base element [7, 8]. This basic element, which made up the matrix, was primarily titanium [9–11], vanadium [12], iron [13], aluminum [14], or nickel [15], amongst others, produced for the aerospace industry with outstanding benefits. According to the conventional alloying system, each element compensates each other's deficiencies to give better properties of the alloy than to give the existence of the materials separately [16, 17]. However, the economy of the fabricated part with the conventional melting approach is the primary drive for research and development of powder metallurgy [18–21].

Landing gear components are typically manufactured with advanced materials using conventional metallic materials such as aluminum alloy 7075 [22–30], alloys steel 4340 [31–36], titanium 6Al-4V [37–42], titanium 6Al-6V-2Sn [43, 44], titanium 15Al-3Cr-3Al-3Sn [45–47], and titanium 10Al-2Fe-3V [48–52]. The most versatile titanium alloys in the aerospace industry are titanium 6Al-4V, titanium 6Al-6V-2Sn, titanium 15Al-3Cr-3Al-3Sn, and titanium 10Al-2Fe-3V because of their excellent properties. In the class of beta titanium alloys, Ti-15Al-3Cr-3Al-3Sn and Ti-10V-2Fe-3Al have been adopted for different airframe applications and landing operations [53]. The alloy Ti-10V-2Fe-3Al has the capability to substitute precipitation-hardening steels due to its in-depth hardenability and exceptional ductility. When it was successfully used in a Boeing 777 aerospace landing gear, the alloy Ti-10V-2Fe-3Al came to limelight. Initially, the alloy cost was higher than that of 4340 steel used in applications for landing gear [35]. However, the cost-benefit of using the alloy was achievable in the long run as the steel parts needed to be

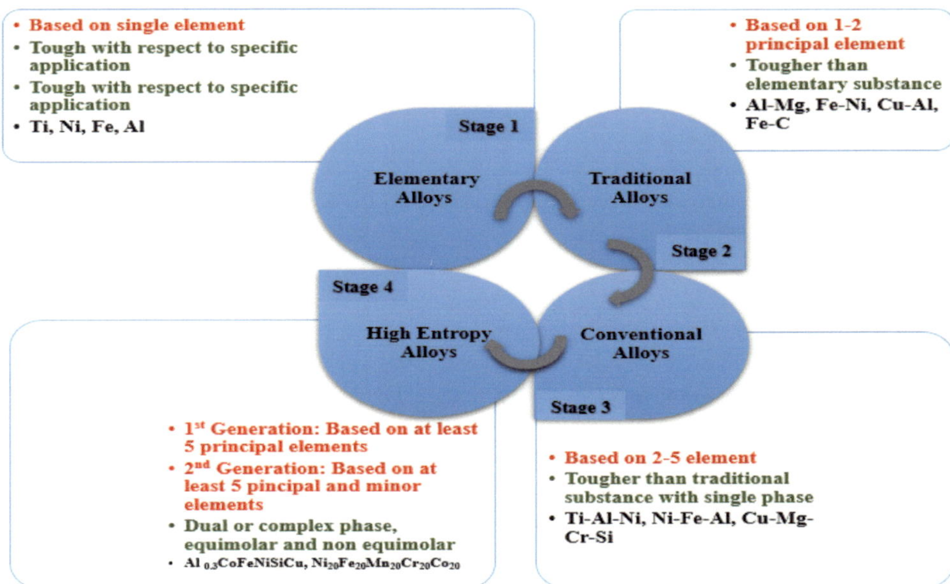

Figure 1.
The evolution and characteristics of alloys.

replaced due to potential susceptibility to stress corrosion cracking [54]. The alloy is therefore applied in many airframe applications of landing gear, wings, fuselage, doors, and cargo handling structures. Similarly, titanium alloys were not impressive for load-bearing components because of uneven stress distribution on parts during operation [55]. The nomenclature of typical landing gear is shown in **Figure 2**, comprising axie, shock, and drag strut, walking beam, and torsion links which are made up of Ti-10V-2Fe-3Al.

Ti-10V-2Fe-3Al popularly used in landing gear and slat/flap tracks was successfully fabricated using alloy forgings [56, 57]. Tracks are the dynamic beams allowing load transmission and movement between wings and moving slats. They are generally produced from maraging steel. Recently, Ti-10V-2Fe-3Al has been made of flap tracks fitted on the trailing side of the wings to perform similar functions. The total weight saving in Boeing 777 was about 40 kg by using six 5 ft. long flap tracks. The assessment of Ti-10V-2Fe-3Al's corrosion resistance is very important given its widespread use as structural parts that are susceptible to wear and corrosion in simulated environments through which an aircraft flies, which also makes the material not exceptional, and surface damage occurs in applications involving contact loading. In regard to this, Arya et al. [58] investigated the electrochemical corrosion performance of Ti-10V-2Fe-3Al alloy in different corrosive media namely, artificial seawater solution, Hank's solution, 0.5 M sulfuric acid, and 0.5 M hydrochloric acid solution. The authors used open circuit potential, current transient time, Tafel extrapolation potentiodynamic polarization curves, and electrochemical impedance spectroscopy to evaluate the corrosion rates. The results show that due to the formation of thick, compact, and stable passive film, this alloy has a lower corrosion rate in alkaline solution. Corrosion rate in acid solution is higher because the passive film is less compact, porous, and unstable. Li et al. [59] used a high-frequency push-pull fatigue testing machine to report the effect of fretting on the fatigue performance of Ti-6Al-4V and Ti-10V-2Fe-3Al alloys. For comparative analysis of the fretting effect on the fatigue performance of the different titanium alloy, the author obtained both plain and fretting fatigue curves. The result shows that Ti6Al4V titanium strength and plain fatigue are lower than Ti-10V-2Fe-3Al titanium strength. But the fretting fatigue of Ti6Al4V titanium is higher under each contact stress. The fatigue source depth of Ti-10V-2Fe-3Al alloy is greater than that of Ti6Al4V alloy. Hardening of Ti-10V-2Fe-3Al alloy is more serious after fretting. The wear mechanism of two titanium alloys is different; Ti1023 titanium alloy is more

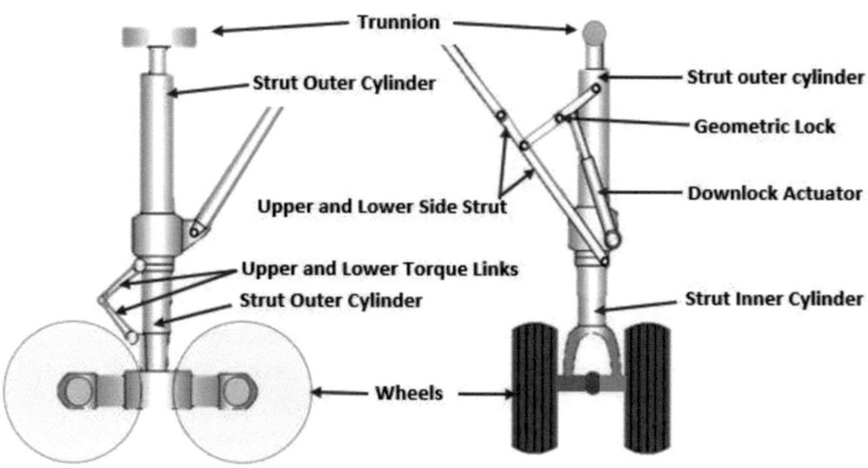

Figure 2.
The nomenclature of main landing gear bogie truck.

sensitive to fretting wear. According to Li et al. [60], the mechanical behavior of the Ti-10V-2Fe-3Al alloy under the high-temperature and dynamic loading conditions and the microscopic observations show that the microstructure of Ti-10V-2Fe-3Al alloy is sensitive to the external thermal loading. As the testing temperature is lower than the β phase transus, the β phase concentration almost remains unchanged besides a few of the refinement and spheroidize of β phase grains, which results in maintaining the high yield strength of the alloy. When the testing temperature is higher than the β phase transus, the thermal loading and mechanical loading can respectively cause various refinements and spheroidize of the β phase grain of alloy, which remarkably decreases the yield strength of the alloy. Based on the resulting stress and strain data obtained in simple shear, the high-temperature dynamic shear deformation mechanisms have been evaluated by using the characteristics of stress-strain behavior caused by thermal loading and mechanical loading. The high-temperature dynamic constitutive model is established based on the experimental results.

Utama et al. [61] reported surface height irregularities that occurred during machining Ti-10V-2Fe-3Al beta (β) titanium alloy. The author observed the height differences in two different regions, "soft region" and "hard region." The hard region has higher Fe and lower Al content, resulting in higher β-phase stability to resist the precipitation of the primary alpha (αp) phase caused by a solution treatment process failure. By contrast, a higher volume fraction of the αp phase and a lower volume fraction of the matrix consisted of a combination of β and secondary alpha (αs) phase.

A high number of αs/β interfaces in the matrix with a predicted 520 HV hardness resulted in harness improvement of the hard region. Hence, the hard and soft regions had different wear resistance capabilities during the machining process, resulting in surface height irregularities. Jiao et al. [62] reported that a complicated and unique phase transformation process takes place with forming different microstructures from forging and reconstruction method to study the spatial distribution of α phase in different grains present in the Ti-10V-2Fe-3Al alloy specimen. Since the process involves repeated heating and cooling during deposition, there was development of the complex and unique phase transformation process in the different microstructures. Their results showed plate-like morphology formation of α laths in the deposited alloy, and a tetrahedral relationship followed their precipitation from the β phase. The different orientation of the grain results in different specific morphologies. The α laths are perpendicular to each other when the cutting plane is parallel to the (100) plane of the BCC cell. If the cutting plane is parallel to the plane (111), a special morphology is established with a mixture of plate-like and rod-like phase α. Liu et al. [63] examined the evolution of microstructure and compression properties of Ti-10V-2Fe-3Al titanium alloy and solution treatments at temperatures ranging from 710 to 830°C, followed by treatment with aging. Ti-10V-2Fe-3Al alloys with α + β-phase have higher mechanical properties than single β-phase alloys. The content of the equiaxed α phase decreases with the increase in solution temperature. As a result, the alloy strength increases as the plasticity decreases. The highest yield strength value of 1668 MPa was obtained in the sample treated with a solution of 770°C treated for 2 hours, then water quenched and 520°C aging for 8 hours followed by air cooling. The stress-induced phase of martensite α" appeared in the sample solution treated at 830°C after dynamic compression of SHPB.

Tabachnikova et al. [64] measured the influence of temperature on Young's modulus for different structural states by mechanical resonance spectroscopy. The heat treatment of the samples led to an increase (~25%) of Young's modulus. Ren et al. [65] established that "Ti-10V-2Fe-3Al has the tendency to spall in a brittle-ductile mixed fracture manner with very rough fracture surfaces covered by ductile dimples". The authors investigated the spall fracture behavior of a typical metastable β-titanium alloy, Ti-10V-2Fe-3Al, by subjecting it to shock loaded through a series of plate impact tests. The microstructural characteristics of the spall planes were systemically investigated for the influence of the shock-induced SIM phase transformation on the fracture behavior of the β-titanium alloy. The authors concluded that the spall strength of Ti-10V-2Fe-3Al is ca. 3.1 GPa at shock pressure below 8.5 GPa dropped by nearly 20% as the shock pressure exceeds 9 GPa due to the abundance of SIM phases with increased shock pressure. The microvoids in the spalled Ti-10V-2Fe-3Al mainly nucleate at the intersections of the grain boundary and the α martensite lath, which obviously showed that the shock-induced β-to-α martensite phase transformation strongly affected the spall fracture behavior of this alloy. The shock-induced β-to-α phase transformation has a detrimental effect on the spallation resistance of titanium alloys, which should be minimized in structural materials of titanium alloys that need to withstand shock wave loading. Srinivasu et al. [53] reported the presence of continuous grain boundary in Ti-10V-2Fe-3Al which is surrounded by a softer precipitate-free zone that leads to flow localization, thereby influencing fatigue crack propagation. The authors therefore studied the effect of processing and heat treatment on the tensile properties and fracture toughness of Ti-10V-2Fe-3Al, a high-strength metastable beta titanium alloy. The alloy was subsequently subjected to thermomechanical processing that included combinations of rolling and solution heat treatment in both α-β and β phase fields. The rolling temperatures ranged from 710°C (sub-transus) to 860°C (super-transus) and the temperatures of the solution treatment ranged from 710°C (sub-transus) to 830°C (super-transus). A systematic microstructural investigation (optical as well as scanning electron microscopy) was undertaken to correlate the property trends with the underlying microstructure, which is strongly dependent on the sequence of thermomechanical processing. A subtransus rolling followed by subtransus solution treatment resulted in a morphology being equiaxed while a super transus rolling followed by subtransus solution treatment resulted in a phase being more acicular/lenticular in morphology. While α-β rolling followed by α-β heat treatment gave better tensile properties, α-β heat treatment followed by β rolling resulted in superior fracture toughness.

In recent years, research has shown that advanced engineering materials used in industries are susceptible to corrosion and fatigue-related failures [66]. Authors attribute these challenges to design shortcomings, poor selection of materials, beyond the design limits, overloading, insufficient maintenance, and manufacturing defects [67]. The resulting repair and replacement costs take up almost half of the total budget; therefore, the performance of materials, particularly fatigue and corrosion resistances, has received widespread attention. According to Hemphill [68], the major drawback of this material is the formation of microcracks, which developed during fabrication as a manufacturing defect causing severe fatigue challenges. Current applications of Ti-10V-2Fe-3Al mainly in the wide range of aircraft structural components [69] are exceeding 200 flying components in each Boeing 777 alone [70, 71]. These applications were made by ingot metallurgy and forging [72], based on Boeing's original development for use in the Boeing 757 [73]. However, a primary problem with the process is the high buy-to-fly ratio [74], which results in about 90% scrap or a material utilization factor of just as much as 10%. Another issue is the severe segregation of α and β phases in the Ti-10V-2Fe-3Al as-cast ingot due to the presence of α and β stabilizers at 15 wt.%. This results in the so-called "beta flecks" in the forged microstructures [69].

Powder metallurgy (PM) allows the manufacture of near-net shape and effectively eliminates segregation by using micrometer-sized powder materials [69, 75].

There are many advantages of powder metallurgy compared to the conventional process in terms of fabricate titanium-based composite, low cost, full densification, no segregation, no inner defect, internal stress, excellent stability, and uniform microstructure. The high buy-to-fly ratio associated with many titanium components, combined with difficulties in forging and machining and recent availability issues, has resulted in a strong drive for near-net titanium manufacturing. PM [76]. is a very promising way to achieve this goal. Accordingly, a variety of PM-based techniques have been implemented to fabricate Ti-10V-2Fe-3Al using either pre-alloyed powder or titanium or TiH2 and master alloy powder blends [71, 77]. Generally, fully dense Ti-10V-2Fe-3Al alloy consolidated by hot isostatic pressing (HIP), or HIP + forging, with or without subsequent solid solution treatment and aging [78], can achieve tensile properties such as wrought alloys. Technically, the simplest and most attractive PM manufacturing route is the cold compaction and sinter approach [69]. Its potential for titanium manufacturing on a variety of titanium materials [79] has been well demonstrated. Although Ti-10V-2Fe-3Al's tensile properties were previously reported [80], the densification process, sintering mechanism, microstructural evolution during sintering, and their correlations with sintered mechanical properties were not given sufficient details. In contrast, the literature on powder-based titanium alloy manufacturing has focused largely on Ti-6Al-4V, with a focus on achieving full densification with respect to different manufacturing conditions [80, 81]. Therefore, there is an ultimate need to understand and establish the fundamentals of the cold-compaction-and-sinter PM approach for the fabrication of high strength or specialty solute rich titanium alloys such as Ti-10V-2Fe-3Al.

Spark plasma sintering (SPS) has been applied in the preparation of alloys, functional materials, ceramics, cermets, nanocomposites, and intermetallic compounds [82]. In recent times, extensive efforts were made towards research and development of SPS as a promising technique for high entropy alloys. Unlike conventional sintering methods, which usually use alternating heat current [83], SPS uses an on-off pulse current to heat the graphite die directly to complete the sintering process. Considering the SPS sintering mechanism, it is specifically assumed that an electrical field is formed between grains during sintering and high-temperature plasmas are excited under the action of pulse current to cause a cleaning effect on the surface of particles, leading to a sintering enhancement [84]. The spark plasma sintering method has several advantages over other conventional sintering techniques like hot pressing, mechanical alloying, and isostatic pressing due to high chances of sintering materials of near full densification and little grain growth [85]. SPS unique features also include a low heating profile, which makes it possible to sinter powders with controlled grain size and limited chemical interaction with other constituents [86]. The SPS method has proven its high effectiveness in fabricating ceramic composites and alloy materials for nano-applications, biomaterials, and electronic materials. From the traditional alloy design concept of titanium 10Al-2Fe-3V, however, HEAs are not based on a single element but on multi-component systems consisting of at least five main elements in an equal or near-equal atomic percentage (at.%) with no noticeable difference between the solute and the solvent [87]. According to existing physical metallurgy and phase diagrams, such multi-element alloys can produce many phases and intermetallic compounds, resulting in complex and fragile microstructures that are difficult to analyze and engineer but are likely to have finite practical values [88]. Experimental results indicate, beyond expectations, that the higher mixing entropy in these alloys enhances the formation of random solid-solution phases with simple structures such as face-centered cubic (FCC) structures [89], body-centered cubic (BCC) structures [90]), or hexagonal close packing (HCP) structures [91], thereby reducing the number of phases [89, 92, 93]. The new alloy design strategy has since opened up an enormous, unexplored multi-component alloy field. The alloy design strategy has achieved unimaginable successes, and great efforts have been devoted to the development and application of many HEAs [94], in various fields because of their excellent performance, such as unique wear resistance [95], excellent strength and thermal stability at elevated temperatures [96], superior high elongation [93], high fatigue, and fracture toughness [97]. Han et al. [98] showed that the "addition of Ti is beneficial to the strength and the compressive ductility of the HEAs at room temperature". The authors presented a new refractory HEA alloy of TiNbMoTaW fabricated from addition of Ti to NbMoTaW alloy. The new TiNbMoTaW HEA was found to have a BBC phase structure that can be maintained even after 24-hour annealing at 1200°C. The room temperature yield strengths and compressive plastic strains of the developed TiNbMoTaW and TiVNbMoTaW HEAs have been significantly enhanced in comparison with the mechanical performance of HEAs NbMoTaW and VNbMoTaW. Also, with impressive yield strengths of ~586 and ~659 MPa at 1200°C, the TiNbMoTaW and TiVNbMoTaW HEAs showed very promising high-temperature strength. As a result of these property profiles, it was suggested that the newly developed TiNbMoTaW and TiVNbMoTaW HEAs have potentials in applications of high-temperature structural materials.

Consequently, extensive research has been conducted to develop next-generation aerospace materials with high mechanical performance and superior corrosion resistance to achieve improvements in both performance and cost repetition. For traditional alloys, one or two principal components are selected based on a specific property requirement, and other alloying components are added to further improve their properties because of the huge formation of bulk intermetallic compounds which occurred as a result of the atomic ratios of elements reaching a 40% mark and above. Thus, this depletes the reliability of the alloy during service. Therefore, the search for an alloy with atomic ratios lower than 35% commenced in 1996 [99]. Hence, the possibility of combining several metallic principal elements in equal atomic compositions was explored and it was named high-entropy alloys (HEAs). HEAs always have the combination of at least five principal elements, each with an atomic percentage (at. %) between 5 and 35%. Generally, the atomic percentage of each minor element, if present at all, is always smaller than 5. The definition is expressed as follows in Eqs. (1) and (2) [94, 102]:

$$n_{major} \geq 5, 5at.\% \leq c_i \leq at35at.\% \tag{1}$$

$$n_{minor} \geq 0, c_j 5at.\% \tag{2}$$

where n_{major} represents the major elements, n_{minor} represents the minor elements, c_i represents the atomic percentages of the major element, and c_j represents the atomic percentages of the minor element.

This definition shows that HEAs need not to be equimolar or near-equimolar, and even contain minor elements to balance various material properties [100]. HEAs can easily be formed into simple FCC phase structures [101], BCC [102], and HCP [91] solid-solution structures. One renowned HEA is an equimolar-shaped Cantor alloy [102] consisting of Fe (BCC), Co (HCP), Cr (BCC), Mn (BCC), and Ni (FCC), with a solid-solution FCC phase when dendritically cooled in the as-shaped sample. Then, by adding Cu (FCC), Nb (FCC), or V (BCC), the five-component alloy was extended to a six-component alloy system, showing the simple solid-solution FCC structure in the as-cast alloy with different lattice parameters. When the HCP type Ti was added, a solid-solution phase of the FCC formed a BCC structure [103]. Another typical example is the AlxCoCrFeNi (molar percentage, $2 \geq x \geq 0$) system prepared by arc melting [104]. The CoCrFeNi as-cast alloy has a solid-solution phase of pure FCC. The AlxCoCrFeNi system changes the crystal structure from FCC to FCC + BCC phases by increasing the Al molar percentage from 0 to 2 and finally to a single BCC phase [104]. A typical TaNbHfZr alloy has a structure with only BCC [105].

Alloy design is a novel means of maximizing and synergizing metal counterparts in HEA multi-material systems. The design encompasses several considerations such as atomic radius, solid solution solubility, powder stoichiometry, processing technique, and the processing conditions or parameters chosen. Thus, HEAs possess dynamic and exceptional properties, such as high hardness and strength profiles, good oxidation and corrosion resistance, and high-temperature softening resistance, which are crucial in prospective engineering applications. HEAs fabricated by SPS have been reported to possess excellent densification properties, as well as high strength and hardness profile. Mohanty et al. [106] research output on a multicomponent equiatomic AlCoCrFeNi high entropy alloy developed by spark plasma sintering established a high mechanical strength. The sintered samples at the optimum temperature of 1273 K displayed the highest microhardness. Fang et al. [85] also researched the mechanical behavior of $Al_{0.5}CrFeNiCo_{0.3}C_{0.2}$ high entropy alloy. They recorded a compressive strength of 2131 MPa and a Vickers microhardness of 617 HV of the $Al_{0.5}CrFeNiCo_{0.3}C_{0.2}$ high entropy alloy. Zhao et al. [107] fabricated HEAs on a carbon steel substrate via spark plasma sintering, which are considered as excellent coating materials due to their high hardness, good wear, and corrosion resistance. The microstructure evolved from FCC to FCC + BCC mixed structure. $Al_xCrFeCoNiCu$ (x = 0, 1, 2, 3) coating has an average hardness of approximately 682 $HV_{0.2}$, which is the highest hardness in all the HEA coatings. Compared with AISI 52100 steel, spark plasma sintered Al2CrFeCoNiCu and Al3Cr-FeCoNiCu HEA coatings show exceptional sliding wear resistance and extremely low friction coefficient in comparison with AISI 52100 steel. Al3CrFeCoNiCu HEA coating wear resistance is approximately four times better than that of bearing steel, showing a promising application as a wear-resistant material According to Chen et al. [108], research outputs of several literature studies indicate that the addition of Ni to Ti-10V-2Fe-3Al is beneficial to the improved mechanical performances of the HEA, making the HEA a good material for high-temperature applications. Nickel is the fifth most common element on earth's crust. As a high-profile element with a good property suited for diverse applications especially that of the aerospace industry, there is no substitute for Ni without reducing performance or increasing cost. The biggest use of nickel is in alloying, particularly with chromium and other metals to produce stainless and heat-resisting steels constituting 65% production [109]. Another 20% is mostly used for highly specialized industrial, aerospace, and military applications. With characteristic melting point of 1453°C, Ni has reliable corrosion and oxidation properties, and readily forms alloys [110, 111].

Chromium, on the other hand, is generally used in metallurgy to impart corrosion resistance, and it also has good strengthening effect of forming stable metal carbides at the grain boundaries of its alloying counterpart [112, 113]. Chromium has excellent mechanical properties such as high corrosion and wear resistance. Therefore, in alloying, it finds a good match with Ni to form Ni-Cr alloys when used with Ti-10V-2Fe-3Al for making landing gear, wind turbine, engine components, and many other industrial and mechanical components where high wear-resistance is needed such as in aerospace applications. The structure of Ni-Cr alloys depends on the percentage composition of nickel or chromium, and the temperature. It is noteworthy that Ni-Cr alloys will be dominated by the π phase, which tends to be brittle at about 60–75% Ni addition. The crystal structure of the FCC is found in the γ phase, and the γ phase shows improved strength and ductility in comparison with the σ phase. The FCC crystal structure is commonly found in Ni-rich alloys, while the BCC crystal structure tends to be found in Cr-rich alloys. The γ phase Ni-Cr alloy can be converted into the ε phase at high pressures, which shows an HCP crystal structure. The alloying element helps stabilize the BCC structure of Ti-10V-2Fe-3Al HEA, increasing the hardness and plastic strain of the alloy. However, there are limited publications in the literature on Ti-10V-2Fe-3Al alloy with Ni and Cr additions, despite the different manufacturing routes that had been established. Moreover, reports investigating the fatigue, corrosion, and oxidation behaviors of spark plasma sintered Ti-10V-2Fe-3Al-Ni-Cr HEAs with their potential applications are scarce. Therefore, this work proposes to use the spark plasma sintering manufacturing technique to synthesize novel Ti-10V-2Fe-3Al-Ni, Ti-10V-2Fe-3Al-Cr, Ti-10V-2Fe-3Al-Ni-Cr, and Ti-V-Fe-Al-Ni-Cr high entropy alloys (HEAs) for making landing gear. Thereafter, the synthesized alloys will be characterized in terms of microstructure, composition, and phase transformation using optical microscopy (OM), X-ray diffraction (XRD), and scanning electron microscopy (SEM) with energy-dispersive X-ray spectroscopy (EDX), respectively. Also, the mechanical and electrochemical properties of the materials will be carried out, and the effects of temperature and stress distributions of the high entropy alloy material during the SPS process will be modeled.

2. Conclusion

From the open literature above, the authors agreed that the current market material (Ti-10V-2Fe-3Al) used in high-temperature applications such as gas turbine and turbine engine in the aerospace industry experience several failures such as high-temperature oxidation and corrosion, limited hardness, and wear resistance. It is confirmed that spark plasma sintering is a potential way to fabricate HEAs which possess properties such as improved microhardness, compressive/tensile strength, tribology, thermal properties, and corrosion resistance properties for low and high-temperature applications as generally agreed by all the authors. Also, it is generally accepted by the authors that SPS processing parameters play a significant role in the mechanical properties of the final developed alloys. The authors concluded that HEAs are the potential replacement for nickel-based superalloys in high-temperature applications. Furthermore, the authors agreed that it is possible to simulate the SPS process by means of finite element modeling. However, SPS simulation of thermal distribution and stress distribution analysis for the development of HEAs are limited in the open literature. Thus, this paper exposes the influence of spark plasma sintering parameters on the mechanical properties of the synthesized alloy.

Author details

Ayodeji Ebenezer Afolabi[1], Abimbola Patricia I. Popoola[1*] and Olawale M. Popoola[2]

1 Department of Chemical, Metallurgical and Materials Engineering, Tshwane University of Technology, Pretoria, South Africa

2 Centre for Energy and Electric Power, Tshwane University of Technology, Pretoria, South Africa

*Address all correspondence to: popoolaap@tut.ac.za

References

[1] Krüger W. Design and simulation of semi-active landing gears for transport aircraft. Mechanics of Structures and Machines. 2002;**30**(4):493-526

[2] Sivaranjani T, Kumar DP, Manjunatha CM, Manjuprasad M. Fatigue life estimation of typical fighter aircraft main landing gear using finite element analysis. In: Advances in Structural Integrity. Singapore: Springer; 2018. pp. 39-52

[3] Saraç H. Shock Absorption Technology for Landing Gears; 2018

[4] Zhang W, Liaw PK, Zhang Y. Science and technology in high-entropy alloys. Science China Materials. 2018;**61**(1):2-22

[5] Adesina O, Popoola P, Fatoba O. Laser surface modification—A focus on the wear degradation of titanium alloy. In: Fiber Laser. IntechOpen.; 2016;**16**:368-381

[6] Dutta S, Robi PS. 5 ChAPtEr Development in materials for sustainable manufacturing. Sustainable Material Forming and Joining. New York: Taylor & Francis Group LLc, London; 2019

[7] Chikumba S, Rao VV. High entropy alloys: Development and applications. In: Proceedings of the 7th International Conference on Latest Trends in Engineering & Technology (ICLTET' 2015); Irene, Pretoria. 2015

[8] Tsao L, Chang S, Yu Y. Direct active soldering of Al 0.3 CrFe 1.5 MnNi 0.5 high entropy alloy to 6061-Al using Sn-Ag-Ti active solder. Transactions of Nonferrous Metals Society of China. 2018;**28**(4):748-756

[9] Jindal P, Worcester F, Walia K, Gupta A, Breedon P. Finite element analysis of titanium alloy-graphene based mandible plate. Computer methods in biomechanics and biomedical engineering. 2019;**22**(3):324-330

[10] Smyrnova KV, Pogrebnjak AD, Kassenova LG. Structural features and properties of biocompatible Ti-based alloys with β-stabilizing elements. In: Advances in Thin Films, Nanostructured Materials, and Coatings. Singapore: Springer; 2019. pp. 319-330

[11] Geetha M, Singh AK, Asokamani R, Gogia AK. Ti based biomaterials, the ultimate choice for orthopaedic implants—A review. Progress in Materials Science. 2009;**54**(3):397-425

[12] Chu Q, Zhang M, Li J, Yan C, Qin Z. Influence of vanadium filler on the properties of titanium and steel TIG welded joints. Journal of Materials Processing Technology. 2017;**240**:293-304

[13] Schinhammer M, Hänzi AC, Löffler JF, Uggowitzer PJ. Design strategy for biodegradable Fe-based alloys for medical applications. Acta Biomaterialia. 2010;**6**(5):1705-1713

[14] Dobrzański LA, Matula G, Dobrzańska-Danikiewicz AD, Malara P, Kremzer M, Tomiczek B, et al. Composite materials infiltrated by aluminium alloys based on porous skeletons from alumina, mullite and titanium produced by powder metallurgy techniques. In: Powder Metallurgy-Fundamentals and Case Studies. IntechOpen.; 2017

[15] Han G, Sato K, Ueno T, Chiba A. Method for manufacturing Ni-based super-heat-resistant alloy. US Patent App. 15/548,447; 2019

[16] Doorbar PJ, Kyle-Henney S. 4.19 Development of Continuously-Reinforced Metal Matrix Composites for Aerospace Applications; 2018

[17] Prasad YV, Rao KP, Sasidhar S, editors. Hot Working Guide: A Compendium of Processing Maps. Ohio, USA: ASM International; 2015

[18] Fang ZZ, Paramore JD, Sun P, Chandran KR, Zhang Y, Xia Y, et al. Powder metallurgy of titanium-past, present, and future. International Materials Reviews. 2018;**63**(7):407-459

[19] Huang R, Riddle M, Graziano D, Warren J, Das S, Nimbalkar S, et al. Energy and emissions saving potential of additive manufacturing: The case of lightweight aircraft components. Journal of Cleaner Production. 2016;**135**:1559-1570

[20] Fang ZZ, Sun P. Pathways to optimize performance/cost ratio of powder metallurgy titanium-a perspective. Key Engineering Materials. 2012;**520**:15-23

[21] Seong S, Younossi O, Goldsmith BW. Titanium: Industrial Base, Price Trends, and Technology Initiatives. Santa Monica, CA: Rand Corporation; 2009

[22] Al-Bahkali EA. Analysis of different designed landing gears for a light aircraft. World Academy of Science, Engineering and Technology. International Journal of Mechanical, Aerospace, Industrial, Mechatronic and Manufacturing Engineering. 2013;**7**(7):1333-1336

[23] Zhu L, Li N, Childs P. Light-weighting in aerospace component and system design. Propulsion and Power Research. 2018;**7**(2):103-119

[24] Kaya H, Uccedil M, Cengiz A, Erguuml RE. The effect of aging on the machinability of AA7075 aluminium alloy. Scientific Research and Essays. 2012;**7**(27):2424-2430

[25] Mahajan Y, Peshwe D. Effect of temper conditions on abrasive Wear behavior of AA7010 alloy. Transactions of the Indian Institute of Metals. 2018;**71**(4):1025-1032

[26] Yildirim M, Özyürek D, Gürü M. The effects of precipitate size on the hardness and wear behaviors of aged 7075 aluminum alloys produced by powder metallurgy route. Arabian Journal for Science and Engineering. 2016;**41**(11):4273-4281

[27] Senthil K, Iqbal M, Chandel P, Gupta N. Study of the constitutive behavior of 7075-T651 aluminum alloy. International Journal of Impact Engineering. 2017;**108**:171-190

[28] Neves RS, Silva DP, Motheo AJ. Corrosion protection of AA7075 aluminium alloy by trimethoxy-silanes self-assembled monolayers. ISRN Electrochemistry. 2013;**2013**:142493

[29] Arriscorreta CA. Statistical Modeling for the Corrosion Fatigue of Aluminum Alloys 7075-T6 and 2024-T3. The University of Utah; 2012

[30] Martín M, Cano M, Castillo G, Herrera M, Martín F. Influence of milling parameters on mechanical properties of AA7075 aluminum under corrosion conditions. Materials. 2018;**11**(9):1751

[31] Yunaidi Y. Pengaruh Jumlah Konsentrasi Larutan Garam Pada Proses Quenching Baja Karbon Sedang S45C. Jurnal Mekanika dan Sistem Termal. 2016;**1**(3):70-76

[32] Ritchie RO, Francis B, Server WL. Evaluation of toughness in AISI 4340 alloy steel austenitized at low and high temperatures. Metallurgical Transactions A. 1976;**7**(6):831-838

[33] Banerjee B. The mechanical threshold stress model for various tempers of AISI 4340 steel. International Journal of Solids and Structures. 2007;**44**(3-4):834-859

[34] Lee W-S, Su T-T. Mechanical properties and microstructural features of AISI 4340 high-strength alloy steel under quenched and tempered conditions. Journal of Materials Processing Technology. 1999;**87**(1-3):198-206

[35] Braga C, dSilva LR, Barbosa EJA, Corrêa ECS. Surface integrity characterization of hardened AISI 4340 steel in grinding process with biodegradable formulations of cutting fluids. Materials Research. 2017;**20**(2):496-501

[36] Zhou J, Retraint D, Sun Z, Kanouté P. Comparative study of the effects of surface mechanical attrition treatment and conventional shot peening on low cycle fatigue of a 316L stainless steel. Surface and Coatings Technology. 2018;**349**:556-566

[37] Ugarte A, M'Saoubi R, Garay A, Arrazola P. Machining behaviour of Ti-6Al-4 V and Ti-5553 Allooys in interrupted cutting with PVD coated cemented carbide. Procedia CIRP. 2012;**1**:202-207

[38] Raghavendra M, Ramachandra C, Srinivas T, Pai MP. Optimization of surface roughness in turning operation in machining of TI-6AL-4V (titanium grade-5). In: IOP Conference Series: Materials Science and Engineering. IOP Publishing; 2018

[39] Antunes RA, Salvador CAF, d Oliveira MCL. Materials selection of optimized titanium alloys for aircraft applications. Materials Research. 2018;**21**(2):1-9

[40] Sterling A, Shamsaei N, Torries B, Thompson SM. Fatigue behaviour of additively manufactured Ti-6Al-4 V. Procedia Engineering. 2015;**133**:576-589

[41] Ritchie RO, Francis B, Server WL. High-cycle fatigue of Ti-6Al-4V. Fatigue and Fracture of Engineering Materials and Structures. 1999;**22**(7):621-632

[42] Razavi S, Bordonaro G, Ferro P, Torgersen J, Berto F. Fatigue behavior of porous Ti-6Al-4V made by laser-engineered net shaping. Materials. 2018;**11**(2):284

[43] Kinney JR. The Power for Flight: NASA's Contributions to Aircraft Propulsion. Vol. 631. Washington, DC: Government Printing Office; 2018

[44] Amateau MF, Kendall EG. A Review of Ti-6A1-6V-2Sn Fatigue Behavior. Aerospace Corp el Segundo ca Materials Sciences Lab; 1970

[45] Maurotto A, Siemers C, Muhammad R, Roy A, Silberschmidt V. Ti alloy with enhanced machinability in UAT turning. Metallurgical and Materials Transactions A. 2014;**45**(6):2768-2775

[46] Demiral M, Nowag K, Roy A, Ghisleni R, Michler J, Silberschmidt VV. Enhanced gradient crystal-plasticity study of size effects in a β-titanium alloy. Modelling and Simulation in Materials Science and Engineering. 2017;**25**(3):035013

[47] Demiral M. Enhanced Gradient Crystal-Plasticity Study of Size Effects in BCC Metal; 2012

[48] Furuhara T, Maki T, Annaka S. Superelasticity in β titanium alloys with nitrogen addition. Journal of Materials Engineering and Performance. 2005;**14**(6):761-764

[49] Quan G, Pu S, Wen H, Zou Z, Zhou J. Quantitative analysis of dynamic softening behaviors induced by dynamic recrystallization for Ti-10V-2Fe-2Al alloy. High Temperature Materials and Processes. 2015;**34**(6):549-561

[50] Yang R, Pan Y, Chen W, Sun Q, Xiao L, Sun J. Deformation behavior

and the mechanism of micro-scale Ti-10V-2Fe-3Al pillarsin compression. Acta Metallurgica Sinica. 2015;**52**(2):135-142

[51] Forged I. Mechanical-Property Data Ti-10V-2Fe-3AI Alloy. Columbus, Ohio: Air Force Wright Aeronautical Laboratory, Materials Laboratory F33615-80-C-5168. 1982

[52] Liu J, Sun J, Chen W. Surface integrity of TB6 titanium alloy after dry milling with solid carbide cutters of different geometriy. The International Journal of Advanced Manufacturing Technology. 2017;**92**(9-12):4183-4198

[53] Srinivasu G, Natraj Y, Bhattacharjee A, Nandy T, Rao GN. Tensile and fracture toughness of high strength β titanium alloy, Ti-10V-2Fe-3Al, as a function of rolling and solution treatment temperatures. Materials & Design. 2013;**47**:323-330

[54] Rocha-Reséndez R, Cabral-Miramontes J, Gaona-Tiburcio C, Zambrano-Robledo P, Estupiñan-López F, Calderon FA. Corrosion behavior of austempered ductile iron used in the aeronautical industry evaluated on acid solutions. In: International Materials Research Congress. Springer; 2016

[55] Cui C, Hu B, Zhao L, Liu S. Titanium alloy production technology, market prospects and industry development. Materials & Design. 2011;**32**(3):1684-1691

[56] Jackson M, Dashwood R, Flower H, Christodoulou L. The microstructural evolution of near beta alloy Ti-10V-2Fe-3Al during subtransus forging. Metallurgical and Materials Transactions A. 2005;**36**(5):1317-1327

[57] Jackson M, Jones N, Dye D, Dashwood R. Effect of initial microstructure on plastic flow behaviour during isothermal forging of Ti-10V-2Fe-3Al. Materials Science and Engineering A. 2009;**501**(1-2):248-254

[58] Arya SB, Bhattacharjee A, Roy M. Electrochemical corrosion behavior of Ti-10V-2Fe-3Al in different corrosive media. Materials and Corrosion. 2018;**69**(8):1025-1038

[59] Li ZY, Liu XL, Wu GQ, Huang Z. Fretting fatigue behavior of Ti-6Al-4V and Ti-10V-2Fe-3Al alloys. Metals and Materials International. 2019;**25**(1):64-70

[60] Li L, Zhang X, Li Z, Wang Z, Ma W. The mechanical behaviours of the Ti-10V-2Fe-3Al alloy under the high-temperature and dynamic loading conditions. In: EPJ Web of Conferences. EDP Sciences; 2018

[61] Utama MI, Ammar AA, Park N, Baek ER. Origin of surface irregularities on Ti-10V-2Fe-3Al Beta titanium alloy. Metals and Materials International. 2018;**24**(2):291-299

[62] Jiao Z, Fu J, Li Z, Cheng X, Tang H, Wang H. The spatial distribution of α phase in laser melting deposition additive manufactured Ti-10V-2Fe-3Al alloy. Materials & Design. 2018;**154**:108-116

[63] Liu AJ, Wang L, Dai HX. Effect of heat treatment on the microstructure and dynamic behavior of Ti-10V-2Fe-3Al alloy. Materials Science Forum. 2018;**910**:155-160

[64] Tabachnikova E, Laktionova M, Semerenko YA, Shumilin S, Podolskiy A, Tikhonovsky M, et al. Mechanical properties of the high-entropy alloy Al0. 5CoCrCuFeNi in various structural states at temperatures of 0.5-300 K. Low Temperature Physics. 2017;**43**(9):1108-1118

[65] Ren Y, Xue Z, Yu X, Tan C, Wang F, Cai H. Spall strength and fracture behavior of Ti–10V–2Fe–3Al alloy

during one-dimensional shock loading. International Journal of Impact Engineering. 2018;**111**:77-84

[66] Xu L, Chen H, Shen L, Che X, Wang Q, Fu Z. Study on fatigue crack propagation behavior in corrosion environment of a cold-rolled austenitic stainless steel. Corrosion. 2017;**73**(8):961-969

[67] Maleque MA, Salit MS. Materials Selection and Design. Singapore: Springer; 2013

[68] Hemphill EA. Fatigue behavior of Al0. 5CoCrCuFeNi high entropy alloys. Acta Materialia. 2012;**60**(16):5723-5734

[69] Yang Y, Luo S, Schaffer G, Qian M. Sintering of Ti–10V–2Fe–3Al and mechanical properties. Materials Science and Engineering A. 2011;**528**(22-23):6719-6726

[70] Boyer RR. Aerospace applications of beta titanium alloys. Journal of Management. 1994;**46**(7):20-23

[71] Welsch G, Boyer R, Collings EW, editors. Materials Properties Handbook: Titanium Alloys. Ohio, USA: ASM International; 1993

[72] Bhattacharjee A, Bhargava S, Varma V, Kamat S, Gogia A. Effect of β grain size on stress induced martensitic transformation in β solution treated Ti–10V–2Fe–3Al alloy. Scripta Materialia. 2005;**53**(2):195-200

[73] Ashley S. Boeing 777 gets a boost from titanium. Mechanical Engineering. 1993;**115**(7):60

[74] Barnes JE, Peter W, Blue CA. Evaluation of low cost titanium alloy products. Materials Science Forum. 2009;**618-619**:165-168

[75] Criswell D. Powder metallurgy in space manufacturing. In: 4th Space Manufacturing; Proceedings of the Fifth Conference. 1980

[76] Boyer RR, Williams JC, Wu X, Clark LP. A realistic approach for qualification of PM applications in the aerospace industry. In: Titanium Powder Metallurgy. Butterworth-Heinemann: Elsevier; 2015. pp. 497-514

[77] Samal P, Newkirk J. Properties and Selection of Powder Metallurgy Titanium and its Alloys Asminternational, Materials Park OH; 2015

[78] Perevoshchikova N, Hutchinson C, Wu X. The design of hot-isostatic pressing schemes for Ti–5Al–5Mo–5V–3Cr (Ti-5553). Materials Science and Engineering A. 2016;**657**:371-382

[79] Luo S, Yan M, Schaffer G, Qian M. Sintering of titanium in vacuum by microwave radiation. Metallurgical and Materials Transactions A. 2011;**42**(8):2466

[80] Agarwal G, Dongare AM. Modeling the thermodynamic behavior and shock response of Ti systems at the atomic scales and the mesoscales. Journal of Materials Science. 2017;**52**:10853-10870

[81] Zhao Y, Li R, Mo J, Tan F, Sun Y. Experimental study on spallation of titanium alloy plates under intense impulse loading. In: Multidisciplinary Digital Publishing Institute Proceedings. 2018

[82] Fu Z, Chen W, Fang S, Zhang D, Xiao H, Zhu D. Alloying behavior and deformation twinning in a CoNiFeCrAl0.6Ti0.4 high entropy alloy processed by spark plasma sintering. Journal of Alloys and Compounds. 2013;**553**:316-323

[83] Pradhan SK, Kalidoss J, Barik R, Sivaiah B, Dhar A, Bajpai S. Development of high density

tungsten based scandate by spark plasma sintering for the application in microwave tube devices. International Journal of Refractory Metals and Hard Materials. 2016;**61**:215-224

[84] Liu Y, Wang J, Fang Q, Liu B, Wu Y, Chen S. Preparation of superfine-grained high entropy alloy by spark plasma sintering gas atomized powder. Intermetallics. 2016;**68**:16-22

[85] Fang S, Chen W, Fu Z. Microstructure and mechanical properties of twinned Al0. 5CrFeNiCo0. 3C0. 2 high entropy alloy processed by mechanical alloying and spark plasma sintering. Materials & Design. 2014;**54**:973-979

[86] Fu Z, Chen W, Wen H, Chen Z, Lavernia EJ. Effects of Co and sintering method on microstructure and mechanical behavior of a high-entropy Al0. 6NiFeCrCo alloy prepared by powder metallurgy. Journal of Alloys and Compounds. 2015;**646**:175-182

[87] Yeh JW, Chen SK, Lin SJ, Gan JY, Chin TS, Shun TT, et al. Nanostructured high-entropy alloys with multiple principal elements: Novel alloy design concepts and outcomes. Advanced Engineering Materials. 2004;**6**(5):299-303

[88] Yeh JW, Chen YL, Lin SJ, Chen SK. High-entropy alloys-a new era of exploitation. Materials Science Forum. 2007;**560**:1-9

[89] Miracle DB, Senkov O. A critical review of high entropy alloys and related concepts. Acta Materialia. 2017;**122**:448-511

[90] Sharma A, Deshmukh SA, Liaw PK, Balasubramanian G. Crystallization kinetics in AlxCrCoFeNi ($0 \leq x \leq 40$) high-entropy alloys. Scripta Materialia. 2017;**141**:54-57

[91] Zhao Y, Qiao J, Ma S, Gao M, Yang H, Chen M, et al. A hexagonal close-packed high-entropy alloy: The effect of entropy. Materials & Design. 2016;**96**:10-15

[92] Miracle D, Senkov O. A Critical Review of High Entropy Alloys and Related Concepts (Postprint). Dayton, OH, United States: UES INC; 2016

[93] Li Z, Tasan CC, Pradeep KG, Raabe D. A TRIP-assisted dual-phase high-entropy alloy: Grain size and phase fraction effects on deformation behavior. Acta Materialia. 2017;**131**:323-335

[94] Pickering E, Jones N. High-entropy alloys: A critical assessment of their founding principles and future prospects. International Materials Reviews. 2016;**61**(3):183-202

[95] Chuang M-H, Tsai M-H, Wang W-R, Lin S-J, Yeh J-W. Microstructure and wear behavior of AlxCo1. 5CrFeNi1. 5Tiy high-entropy alloys. Acta Materialia. 2011;**59**(16):6308-6317

[96] Zou Y, Ma H, Spolenak R. Ultrastrong ductile and stable high-entropy alloys at small scales. Nature Communications. 2015;**6**:7748

[97] Gludovatz B, Hohenwarter A, Catoor D, Chang EH, George EP, Ritchie RO. A fracture-resistant high-entropy alloy for cryogenic applications. Science. 2014;**345**(6201):1153-1158

[98] Han Z, Chen N, Zhao S, Fan L, Yang G, Shao Y, et al. Effect of Ti additions on mechanical properties of NbMoTaW and VNbMoTaW refractory high entropy alloys. Intermetallics. 2017;**84**:153-157

[99] Aristeidakis IS, Tzini MIT. High Entropy Alloys; 2016

[100] He JY, Wang H, Huang HL, Xu XD, Chen MW, Wu Y, et al. A precipitation-hardened high-entropy alloy with outstanding

tensile properties. Acta Materialia. 2016;**102**:187-196

[101] Manivasagam G, Suwas S. Biodegradable Mg and Mg based alloys for biomedical implants. Materials Science and Technology. 2014;**30**:515-520

[102] Zhang Y, Zuo TT, Tang Z, Gao MC, Dahmen KA, Liaw PK, et al. Microstructures and properties of high-entropy alloys. Progress in Materials Science. 2014;**61**:1-93

[103] Cantor B, Chang ITH, Knight P, Vincent AJB. Microstructural development in equiatomic multicomponent alloys. Materials Science and Engineering. 2004:213-218

[104] Wang WR, Wang WL, Wang SC, Tsai YC, Lai CH, Yeh JW. Effects of Al addition on the microstructure and mechanical property of AlxCoCrFeNi high-entropy alloys. Intermetallics. 2012;**26**:44-51

[105] Maiti S, Steurer W. Structural-disorder and its effect on mechanical properties in single-phase TaNbHfZr high-entropy alloy. Acta Materialia. 2016;**106**:87-97

[106] Mohanty S, Maity T, Mukhopadhyay S, Sarkar S, Gurao N, Bhowmick S, et al. Powder metallurgical processing of equiatomic AlCoCrFeNi high entropy alloy: Microstructure and mechanical properties. Materials Science and Engineering A. 2017;**679**:299-313

[107] Zhao Y, Yang Y, Lee C-H, Xiong W. Investigation on phase stability of AlxCo0.2Cr0.2Ni0.2Ti0.4−x high entropy alloys. Journal of Phase Equilibria and Diffusion. 2018;**39**(5):610-622

[108] Chen M-R, Lin S-J, Yeh J-W, Chen S-K, Huang Y-S, Tu C-P. Microstructure and properties of Al0. 5CoCrCuFeNiTix (x = 0-2.0) high-entropy alloys. Materials Transactions. 2006;**47**(5):1395-1401

[109] Davis JR. Alloying: Understanding the Basics. Ohio, USA: ASM International; 2001

[110] Tundermann JH, Tien JK, Howson TE, UB Staff. Nickel and nickel alloys. Kirk-Othmer Encyclopedia of Chemical Technology. 2000:1-19

[111] Bide T, Hetherington L, Gunn G. Nickel. British Geological Survey; 2008

[112] Moutarlier V, Neveu B, Gigandet MP. Evolution of corrosion protection for sol-gel coatings doped with inorganic inhibitors. Surface and Coatings Technology. 2008;**202**(10):2052-2058

[113] Tong C. Advanced materials enable energy production from fossil fuels. In: Introduction to Materials for Advanced Energy Systems. Springer; 2019. pp. 171-230

Chapter 10

Cutting Superhard Materials by Jet Methods (on Functional Approach)

Alexandr Salenko, Viktor Shchetynin, Galina Gabuzian, Evgeny Lashko, Mohamed R.F. Budar, Sergey Klimenko and Alexandr Potapov

Abstract

Dimensional processing of products made from hard alloys and superhard materials is of a certain interest for contemporary high-technology production, as it enables creation of half-finished or final products of various geometric shapes from workpieces obtained by sintering. In this case conventional methods, first of all abrasive processing and electroerosion cutting, find limited application, which is caused by special features of the structural condition and physical and mechanical properties of the materials of the processed workpieces. The functional approach can be used in working of composite workpieces from laminated superhard materials. The performed research with the use of hydro-abrasive cutting (HAC), laser cutting (LC), laser cutting with water cooling (LCC), and water jet-guided laser (WJGL) has enabled studying of the intensity of destruction area introduction into the worked piece and finding out the functional features of a particular process that are caused by the working conditions. It has been shown that the hypothesis of quasistationary rate of destruction is unacceptable for such materials, which is caused by the structure and high hardness of the material. It has also been determined that when deepening increases, the rate of jet introduction has a pronounced tendency toward reduction.

Keywords: carbon materials, carbon composite, water-jet-guided laser process, jet methods

1. Introduction

Dimensional processing of products made from hard alloys and superhard materials, in particular, contour cutting prismatic and circular plates and arbitrary shaped flat workpieces, is of a certain interest for contemporary high-technology production, as it enables creation of half-finished or final products of various geometric shapes from workpieces obtained by sintering.

In this case conventional methods, first of all abrasive processing and electroerosion cutting, find limited application, which is caused by special features of the structural condition and physical and mechanical properties of the materials of the processed workpieces. Processing of products made from laminated compounds is especially complicated as there is a danger of breakage of adhesion bonds in the "base-surface layer" plane. Workpieces made from such products include diamond carbide (DC) composite containing an upper (working) layer of polycrystal superhard composite on the basis of synthetic diamond (PCD) and a lower (supporting) layer from hard alloy based on tungsten carbide (HA), obtained by sintering in high-pressure apparatus [1].

It was demonstrated that, first of all, liquid blasting and laser blasting have good prospects in industrial use for cutting flat workpieces from polycrystal superhard materials (PSHM) and HA, as they allow creation of the cut surface with sufficient efficiency. However, the problems of provision of high quality of the worked surfaces of the products have not yet been solved in full.

So, surfaces of products from HA and PSHM obtained by the mentioned methods are characterized by high roughness and essential deviation of the shape. In Ref. [2], it is shown that hybrid processing methods based on a combination of different ways of power and other flows impact on the material that make it possible to essentially reduce working hours necessary for production and improve the quality of processing. At the same time, analysis of the final product from the point of view of its useful functions enables improvement of the method and scientific substantiation of most rational ways of impact on the workpiece to achieve the maximum quality level.

The solution to these problems can be found in the use of an innovative approach to development of hybrid working methods; its essence consists in provision of useful functions of the product on the basis of morphological analysis of variants of combination of power and energy flows generating a hybrid production process.

The purpose of the chapter consists in development of principles of functional approach to creation of hybrid processes as a morphological combination of various ways of power and energy flows that impact on the worked piece when functions and properties of the final product are formed by totality of results of some technological transitions realized on micro-, meso-, or macro-levels of the product.

2. Approaches to problem solution and research methods

Functional approach to development and creation of new machines, objects, and complex technical systems was studied by many researchers [3]. They state that any material object is characterized by a certain totality (matrix) of functions among which it is possible to single out useful, harmful, and neutral functions (**Figure 1**). Unlike a material approach, a functional approach is based on the fact that the product is made to perform a number of functions provided by corresponding material carriers (the cheapest ones or the ones with the least costly manufacturing steps).

This approach can also be applied to working technologies: the manufacturing process expressed through material carriers is to be minimized according to criteria taken into consideration—working time, cost price, and quality. Systemized data of this approach are presented in a number of papers, for example, [4].

The idea of modularity of technological processes (TP) and their functional orientation can develop in the following direction.

As all types of functions (useful, neutral, and harmful) are available in a final product, manufacturing steps are to be oriented in such a way that harmful

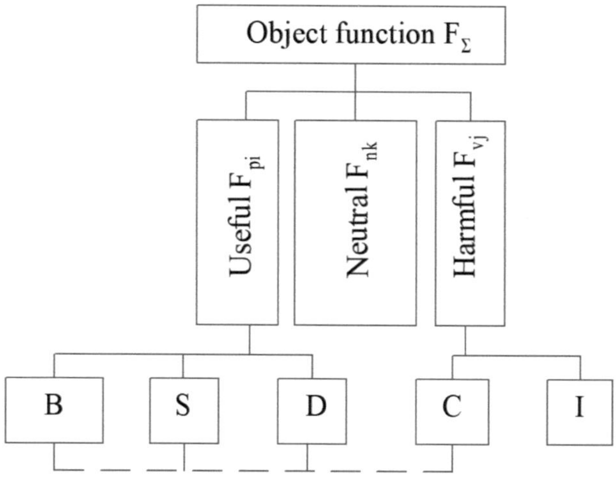

Figure 1.
Classification of the object functions: B, basic; S, secondary; D, derived (obtained without special provision); C, connected (accompany useful functions); I, independent.

functions be weakened and useful ones, in their turn, be obtained in the minimum number of steps. Under these conditions a technological process can be considered prospective if weakening or complete elimination of harmful functions takes place along with creation of useful functions during the steps.

Analysis of typical products of mechanical engineering from the point of view of functional approach reveals that practically always creation of a particular useful consumer function F_p will go together with manifestation of neutral F_n and harmful F_v functions. Then a product having only useful (under certain conditions) functions is ideal from the point of view of operation:

$$F_p = F_{pz}, F_v = 0, F_n \to \min, \tag{1}$$

where F_{pz} are the product useful functions having the following matrix of consumer properties

$$\boldsymbol{P} = \sum_{i=1}^{l} F_{pi} + \sum_{j=1}^{m} F_{nj} + \sum_{k=1}^{p} F_{vk}. \tag{2}$$

Hence, an equation of restrictions (1) and optimization (2) makes it possible to choose the most rational material carriers of functions on the basis of morphological analysis and then to pass to material carriers in the technological process creating these functions.

As there is a functional interrelation between separate functions, that is,

$$F_v = pF_p, F_n = qF_p. \tag{3}$$

taking into account the fact that a function is created by a separate TP step in the form of transformation element W_p, Eq. (2) can be presented in the following form:

$$\boldsymbol{P} = \sum_{i=1}^{l} W_{\mathrm{p}} F_{pi} + \sum_{j=1}^{m} W_n q F_{pj} + \sum_{k=1}^{p} W_v p F_{pk}. \tag{4}$$

Taking into consideration the fact that TP cannot be aimed at creation of harmful functions and properties in the product, minimization of the unnecessary functions is expressed by dependence:

$$\boldsymbol{P} = \sum_{i=1}^{l} W_{\mathrm{p}} F_{pi} + \sum_{j=1}^{m} \overline{W_p q} F_{pj} + \sum_{k=1}^{p} \overline{W_p p} F_{pk}, \tag{5}$$

or after transformation, in the following way

$$\boldsymbol{P} = \sum_{i=1}^{l} F_{pi} \left(W_p + \overline{W_p q} + \overline{W_p p} \right). \tag{6}$$

Approaching of the totality of the product properties to the ideal implies transformation of summands $\overline{W_n q}$ and $\overline{W_v p}$ into zero, which is possible under the condition of absence of functional relation between useful and harmful properties of the product or under the condition that the process of obtaining useful properties due to a particular TP step is at the same time the inverse one as to the occurring harmful functions. Availability and interrelation of TP separate elements enable presentation of Eq. (6) in the following way:

$$\boldsymbol{P} = \sum_{i=1}^{l} F_{pi} \left(W_p + \overline{W_p q} + \overline{W_p p} + \overline{W_p} (p + q) \right). \tag{7}$$

For totality of alternative variants of TP, the obtained equality is supplemented by quantitative signs of every function, the integral sum of which is equal to 1, then

$$I = \sum_{i=1}^{l} F_{pi} \left(b_{ki} W_{pi} - b_{vi} \overline{W_{pi} p} - b_{ri} \overline{W_{pi}} (p + q) \right). \tag{8}$$

under the condition that $\overline{b_{ni} W_{pi} q} = 0$, where b_{ki} and b_{vi} are the corresponding weight coefficients of each of the useful and harmful functions and b_{ri} is the weight coefficient of interaction of independent steps revealing reserves in improvement of output properties of the final product.

Describing the object by setting its initial condition C_n, as a totality of parameters characterizing the form and dimensions of the workpiece, its physical and mechanical properties, and final condition C_k via particular forms (dimensions, relative position of the surfaces, physical and mechanical properties, etc.), the technological transformation function φ_0 is presented as

$$\varphi_0 : \left\{ \begin{array}{c} C_{n1} \\ C_{n2} \\ \vdots \\ C_{nR} \end{array} \right\} \rightarrow \left\{ \begin{array}{c} C_{k1} \\ C_{k2} \\ \vdots \\ C_{kT} \end{array} \right\} \tag{9}$$

where C_{nR} is r-th elementary property of the workpiece; C_{kT} is t-th elementary property of the product; and R and T are total number of parameters of the workpiece and the product, respectively. The function $\varphi_0 = \Sigma W_i E_j$, where E_j is the product separate elements creating its properties.

As the product separate functions expressed via obtaining parameters of geometric accuracy, condition, structure, etc. can be generated in different ways, it is expedient to create morphological tables that may provide the basis for search of more rational variants of combination of technological actions.

Let the created product realize some totality of functions $F_1...F_l$. To guarantee them the elements of condition of workpiece, E_{ki} is to be formed in such a way (**Figure 2**) that

$$F_l = C_{ki} = C_{pi}W_{pj} = \Sigma\, E_{ki}W_{pj}.$$

Application of this approach makes it possible to reveal rational sequence of TP steps, find causes of occurrence of harmful functions and eliminate them (or invert), and also determine the possibility for hybridization of the process to meet condition (1).

As generation of assigned functions is a multivariant task (**Figure 2** and **Table 1**), logical relations obtained on the basis of morphological analysis can be simplified according to the known rules of Boolean algebra provided that restrictions be met (7). In this case differentiation or hybridization of operation may have new effects that are to be taken into account by corresponding weight coefficients (8).

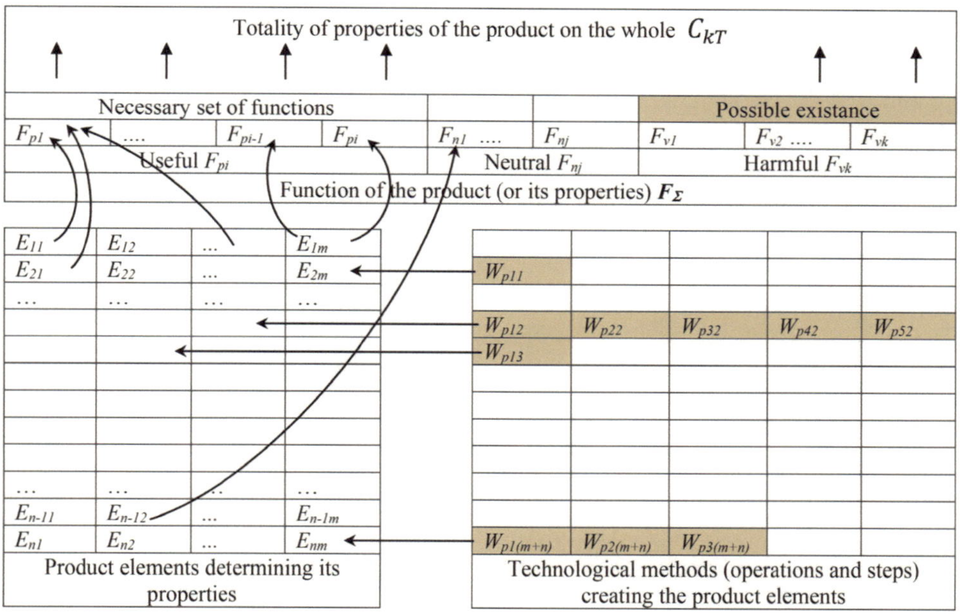

Figure 2.
Provision of the product functions F_{pi} by separate elements when they are generated by TP steps W_{ij}.

No. п/п	Product functions	Element provision	Material carriers—TP steps	Probable harmful functions
1	2	3	4	5
1	F_{p1}	$E_2 \cap E_3 \cap E_4$	$W_{p1} * W_{pj} + W_{p3}$	—
2	F_{p2}	$E_2 \cap E_3 + E_7 \cap E_8$	$W_{p4} * W_{p7} + W_{p3} * W_{p5} + W_{p2}$	F_{v1}
...				
L	F_{pl}			

Table 1.
Components of the product useful functions and their provision by TP.

The product properties are generated as a result of a number of manufacturing steps during which a complete or partial change of the initial properties takes place. Technological transformation of a workpiece into a product is achieved by purposeful total technological impacts $W_{ij}(t_k)$ of material $S_o(t_k)$, energy $E_o(t_k)$, and information $I_o(t_K)$ types which enable presentation of a scheme of output properties generation according to **Figure 3** and write down.

$$W_{ij}(t_k) = S_o(t_k) \cup E_0(t_k) \cup I_o(t_k).$$

Then, on the grounds of the condition that tool technological impacts on the product are to be performed at the levels from nano-areas to the product on the whole, and the product is a 3D object, to realize the totality of variants of technological impacts the morphological matrix will correspond to the following form:

$$A_3^{\Pi} = \begin{Vmatrix} \Pi_{11}^S & \Pi_{12}^S \dots & \Pi_{21}^S & \Pi_{22}^S \dots & H_{11}^S & H_{12}^S \dots & H_{21}^S & H_{22}^s \dots & E_{11}^S & E_{12}^S \dots & E_{21}^S & E_{22}^S \dots \\ \Pi_{11}^t & \Pi_{12}^t \dots & \Pi_{21}^t & \Pi_{22}^t \dots & H_{11}^t & H_{12}^t \dots & H_{21}^t & H_{22}^t \dots & E_{11}^t & E_{12}^t \dots & E_{21}^t & E_{22}^t \dots \\ \Pi_{11}^v & \Pi_{12}^v \dots & \Pi_{21}^v & \Pi_{22}^v & H_{11}^v & H_{12}^v \dots & H_{21}^v & H_{22}^v \dots & E_{11}^v & E_{12}^v \dots & E_{21}^v & E_{21}^v \dots \end{Vmatrix}$$

where $\Pi_{11}^s, \Pi_{12}^s, \dots; \Pi_{21}^s, \Pi_{22}^s, \dots;$ $\Pi_{11}^t, \Pi_{12}^t, \dots; \Pi_{21}^t, \Pi_{22}^t, \dots; \dots \Pi_{11}^v, \Pi_{12}^v, \dots; \Pi_{21}^v, \Pi_{22}^v, \dots; \dots$—variants of discontinuous technological actions along the corresponding axes s, t and v of the coordinate system of s, t, v; $H_{11}^s, H_{12}^s, \dots; H_{21}^s, H_{22}^s, \dots;$. $H_{11}^t, H_{12}^t, \dots; H_{21}^t, H_{22}^t, \dots;$ $H_{11}^v, H_{12}^v, \dots; H_{21}^v, H_{22}^v, \dots$—different variants of continuous technological actions along the axes s, t and v of the coordinate system of s, t, v; $E_{11}^s, E_{12}^s, \dots; E_{21}^s, E_{22}^s, \dots;$—different variants of one-time technological actions.

The presence of variants of discontinuous technological actions makes it possible to consider processing of one element of the product in the form of a successive

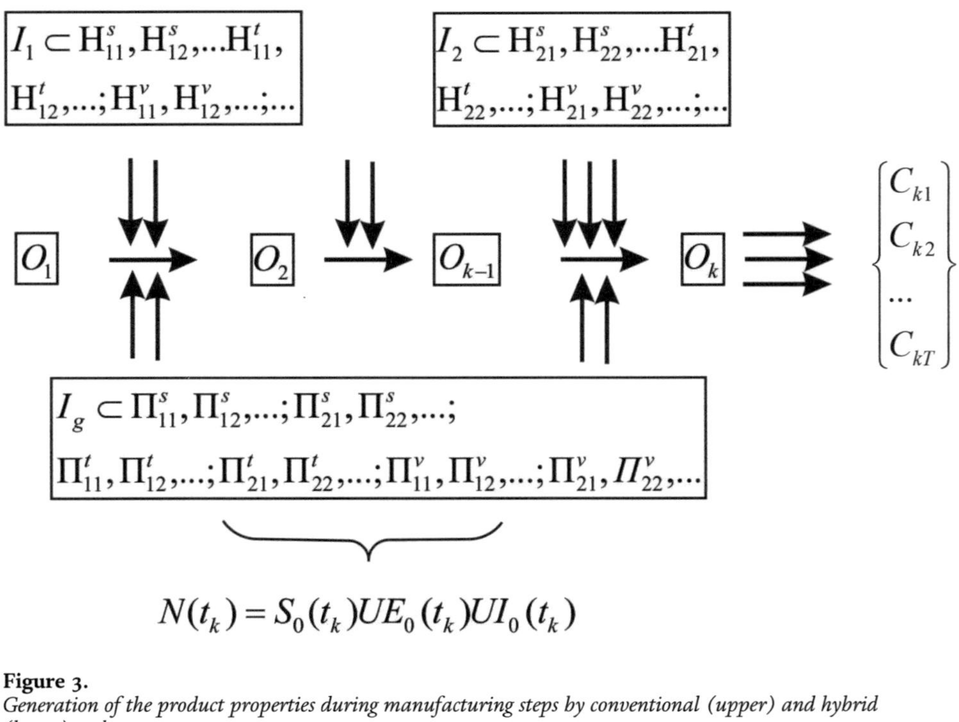

Figure 3.
Generation of the product properties during manufacturing steps by conventional (upper) and hybrid (lower) tools.

totality of different actions. In this case if the element geometric characteristics (e.g., flatness, accuracy of linear dimensions) are its output index, this process can be realized by different types of actions that more completely correspond to the properties of the workpiece elements. As actions providing the conditions for minimum error of the shape are to be taken without reinstallation of the workpiece and change of its position in the fixation and orientation device, such processes should be considered hybrid ones.

Let some element E_m of a product be obtained due to realization of discontinuous technological actions Π^t_{ij} and Π^t_{ij+k}. It can be expected that k tools, accordingly, will be required for the realization. However, if it is taken into account that a new tool is created on the basis of the known ones, that is, expression $R_{nj} = \bigcap_{i=1}^{\rho_i} R_{si}$ takes place, where R_{nj} is the field of creation of new types of tools; R_{si} is the i-th totality of the known engineering solutions; and ρ_i is the weight of the subset of the known engineering solutions, then the newly created tool can combine the means for fundamentally different types of actions (**Figure 3**).

Let totalities of properties of two tools represent expressions

$$I_1 = \begin{Bmatrix} \rho_{11}{S^1}_{11} & \rho_{21}{S^1}_{21} & \rho_{k1}{S^1}_{k1} \\ \cdots & \cdots & \cdots \\ \rho_{1j}{S^1}_{1j} & \rho_{2j}{S^1}_{2j} & \rho_{kj}{S^1}_{kj} \end{Bmatrix}, \; I_2 = \begin{Bmatrix} \rho_{11}{S^2}_{11} & \rho_{21}{S^2}_{21} & \rho_{k1}{S^2}_{k1} \\ \cdots & \cdots & \cdots \\ \rho_{1j}{S^2}_{1j} & \rho_{2j}{S^2}_{2j} & \rho_{kj}{S^2}_{kj} \end{Bmatrix}. \quad (10)$$

Then a hybrid tool obtained on the basis of the principle of morphological search and combination of properties will consist of m elements and $m < k + j$, as some of properties of initial tools can be combined. Thus, the index of hybridization of the created instrument is $k_g = \frac{k+j}{m}$. This index provides the possibility to find rational engineering solutions for a hybrid tool on the grounds of a totality of the required properties of the worked product and also the possibilities to achieve them by available means.

Consider specific features of application of this approach to creation of hybrid processes for working of diamond-bearing products intended to be used as tools. Such products are rather simple, the number of elements E_i, defining functions F_j of the latter is not large and, as a rule represents several surfaces making working surfaces of the tool and a cutting element fastening plane.

Conventionally, these products represent plates of various geometric forms and are made homogeneous or laminated depending on their purpose. They are mainly worked up by an abrasive tool (AT). The use of various abrasive wheels enables obtaining flat elements of surface (during cutting or profile ones during wheel periphery copying (**Figure 4a**).

An alternative consists in application of laser working methods, in particular, laser cutting (LC), laser cutting with water cooling (LCC) or water-jet-guided laser (WJGL), and also a method using loose abrasive accelerated by a supersonic liquid flow—hydro-abrasive cutting (HAC). In this case the obtained surfaces can be of an arbitrary form determined by operational movement of the tool in relation to the worked surface (**Figure 4c**).

Peculiar features of application of these processes are discussed in [1, 3, 5].

As preliminary research has revealed, separate application of these methods is not optimal.

Let us consider specific features of LC, LCC, WJGL, and HAC from the point of view of a functional approach.

HAC is known to generate a stressed condition at the obstacle—the worked surface; this state is described by components

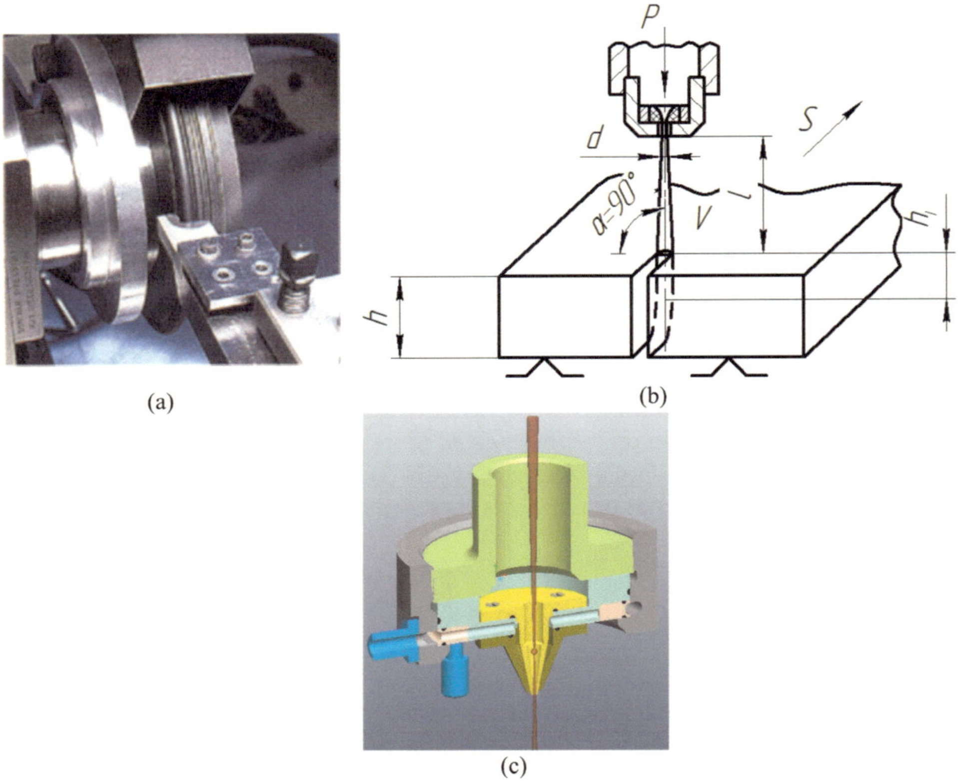

Figure 4.
Processing tool working surface shaped abrasive wheel (a), model of a jet device (b), a laser head for cutting the model (c).

$$\sigma_r = 2G\left(\frac{\partial U}{\partial r} + \frac{\mu\varepsilon}{1-2\mu}\right); \sigma_t = 2G\left(\frac{U}{r} + \frac{\mu\varepsilon}{1-2\mu}\right);$$
$$\sigma_z = 2G\left(\frac{\partial H}{\partial z} + \frac{\mu\varepsilon}{1-2\mu}\right); \tau = 2G\left(\frac{\partial H}{\partial z} + \frac{\mu\varepsilon}{1-2\mu}\right);$$
$$\varepsilon = \varepsilon_r + \varepsilon_t + \varepsilon_z = \frac{\partial U}{\partial r} + \frac{U}{r}\frac{\partial H}{\partial z}; \varepsilon = \frac{1-2\nu}{2(1-\nu)G}(\sigma_r + \sigma_t + \sigma_z)$$
$$= \frac{1-2\nu}{E}(\sigma_r + \sigma_t + \sigma_z)\begin{cases} (1-2\nu)\left[\Delta U - \frac{U}{r^2}\right] + \frac{\partial\varepsilon}{\partial r} = 0 \\ (1-2\nu)\Delta U + \frac{\partial\varepsilon}{\partial r} = 0. \end{cases},$$

where $U(t)$ and $H(t)$ are components of movements at a particular point of the surface that are determined as $U(t) = -\frac{(1-2\nu)p_o(t)\left[\frac{D_k}{2}\right]^2}{4G}; 0 < \frac{D_i}{2} < \frac{D_k}{2};$ $H(t) = -\frac{(1-2\nu)p_o(t)D_k}{2G}$; G and μ are the shift module and Poisson ratio of the worked material, respectively; ε is the volume deformation; $\Delta = \frac{d^2}{dr^2} + \frac{d}{rdr} + \frac{d^2}{dz^2}$ is the Laplace operator; and p_O is the pressure at the obstacle $p_i = 0,5\rho v_i^2 + \rho v_i^2 dQ$.

Intensification of the stressed condition contributes to development of initial defects and creation of a grid of microcracks actively joining under the action of abrasive particles when material particles come off the surface. HAC is characterized by the property to efficiently continue working only till the moment when the loss of jet energy due to friction against the surface of the funnel that appeared is comparable with the energy at which the cutting process ceases.

In particular, in [1, 5] it is shown that during working of carbides (HA), the dimple depth h_l increases due to manifestation of mechanisms of micro-cutting and deformation destruction, that is, the process is described by the expression $h_l = c\sqrt{\frac{m_a v^2}{8\sigma u^2} + \frac{2m_a(1-c)v^2}{\pi u \varepsilon D_j}}$, where c is the process constant; m_a is the abrasive mass flow; v is the velocity of abrasive particles movement, $v = \frac{2p_b f_b}{f_b\sqrt{2p_b/\rho + m_a}}$; σ is the material flow stress; u is the feed rate; ε is the material specific energy; and D_j is the flow diameter corresponding to the diameter of the nozzle section.

On the other hand, due to flow energy losses, the real depth of the obtained dimple will be less and can be taken into account by the relation

$$h_l = c\sqrt{\frac{m_a v^2}{8\sigma u} + \frac{(1-N_1)^2 d_j}{\frac{N_2 N_3}{1-c} + C_f(1-N_1)}}, \tag{11}$$

where $N_1 = \frac{v_c}{v}$; $N_2 = \frac{\pi}{2}\frac{\varepsilon d_j^2}{m_a v}$; $N_3 = \frac{u}{v}$; v_c is the critical velocity of abrasive particles movement.

A peculiar feature of cutting hard composite workpieces used in tools both separately and in the form of bases for the required layers of other materials consists in the fact that particles flowing on the surface cause local highly intensive loads resulting in some elastoplastic compressive macrodeformations in local volumes of the surface layer. These loads are mainly received by the carbide structure (for alloys containing cobalt ≤10%). Further pickup of the abrasive particle by liquid flow results in removal of compression load and partial elastic restoration of the deformed volume of the surface layer, that is, in occurrence of tensions in this local volume, which causes redistribution of tensions among the components of HA structure. In this case, at first carbide grain boundaries break, which results in appearance of microcracks in HA carbide grains themselves and plastic deformation along the dislocation mechanism of cobalt bundle [2, 3]. After that the boundaries between the carbide grains and the bundle and the bundle itself break.

Hence, movement of the destruction area slows down when the cutting depth increases, and distortion of the jet as a "nonrigid" tool increases.

Much higher rigidity of polycrystal superhard materials, including cubic boron nitride (PSHM, CBN) and diamond-bearing elements (DBE), causes the fact that a moving flow of abrasive grains slightly influences the worked material and the workpiece is not cut at a speed admissible by manufacturing conditions (**Figure 5a**).

Obtaining the initial groove and creation of a cut or a channel result in distortion of the jet and its drift from the theoretic axis to the side opposite to the feed movement. It is facilitated by selectivity of destruction caused by "nonrigidity" of the jet, due to which creation of specific destruction areas takes place (**Figure 5b**); their location and dimension determine the form of hydro-cutting front and deviation of the jet by angle α, whose value is determined by the relation of speed v_z of jet penetration into the worked material and feed rate s. This phenomenon results in appearance of surface defects in the form of waviness and also in cut edge deviation from orthogonality.

The jet ability to selectively go about obstacles results in the fact that heterogeneity of the worked material and specific features of the power scheme may provoke availability of both bumpy elements of the surface and surface cleavages (**Figure 5c**).

Thus, the possibilities of HAC process from the point of view of the functional approach can be presented by elements of **Table 2**. In this case the ability of the jet to destroy the obstacle with creation of vertical edges can be regarded as a useful function only for a restricted number of materials.

Then

$$W_{1i}(t_k) = W_1^{Fp1}(t_k) \cap W_1^{Fv1}(t_k) \tag{12}$$

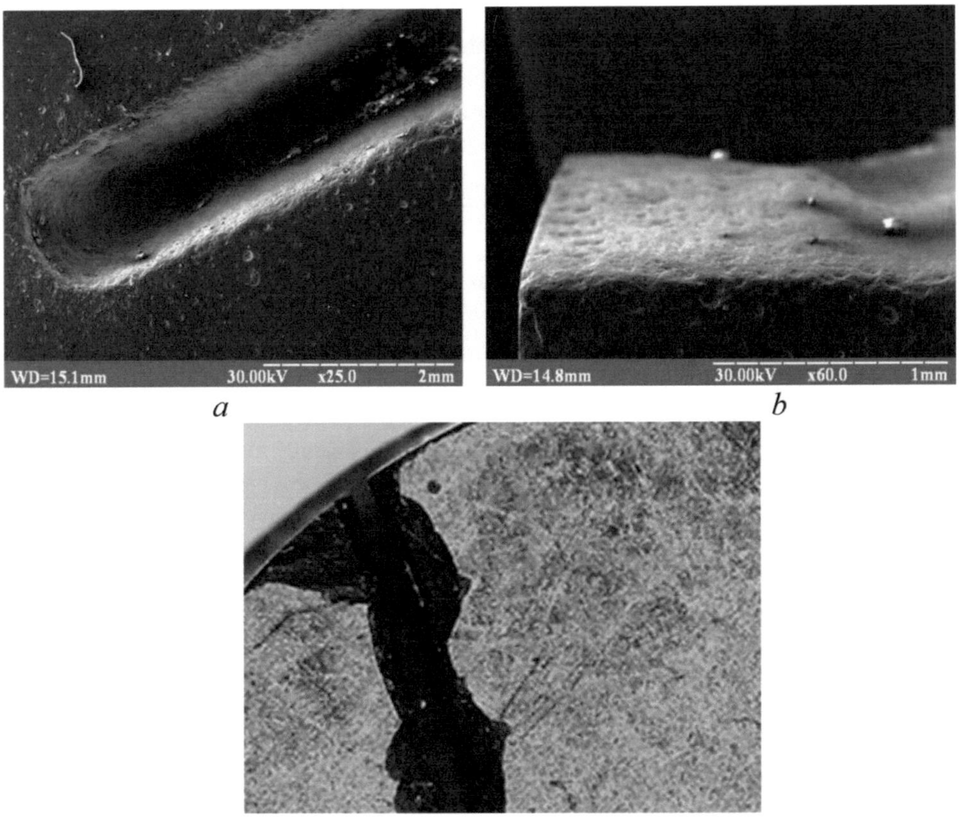

c

Figure 5.
Surface defects as manifestation of harmful functions of technological actions: (a) incomplete cutting; (b) surface waviness; (c) surface cleavages.

Worked material	Functions F'			
	Useful F_{pi}	**Harmful F_{vk}**		**Neutral F_{nj}**
	Obtaining of orthogonal edge of the product h, mm	**Waviness δ, μm**	**Surface cleavage**	**Workpiece heat, T (°C)**
HA	F_{p1} depends on cutting modes	F_{vk} depends on cutting modes	Minimum	Up to 50–60
PSHM	Weak	High	Complete	Up to 50–60
PCD	Absent	Absent	Insignificant	Up to 50–75

Table 2.
Provision of the product function by creation of element E_i by a technological action $W_{ij}(t_k)$ at HAC.

under the condition that F_{v2} and F_{n1} can be neglected and useful and harmful properties (functions) are manifested simultaneously. Corresponding transformations in the form of the process result, for example, depth h (the surface element linear value) for HA, may represent regression equations in the form

$$h|_{W^{Fp}} = b_0 + b_1 s_k + b_2 M_a;$$
$$\delta|_{W^{Fv}} = b_0 + b_1 s_k + b_2 h_m, \tag{13}$$

where M_a is the abrasive grains mass flow, h_m is the thickness of the worked material, and for other materials $h|_{W^{Fp}} = 0$ and $\delta|_{W^{Fv}} = 0$.

It is obvious from the given relations that these two dependences are interconnected and increase of cutting depth h at a higher rate of contour feed s_k results in increase of waviness δ.

More possibilities are provided by combined working of material by wate-jet-guided laser (WJGL). In this case working variants correspond to schemes in **Figure 6**; in this case both WJGL (**Figure 6a**) and LCC (**Figure 6b–e**) can be realized.

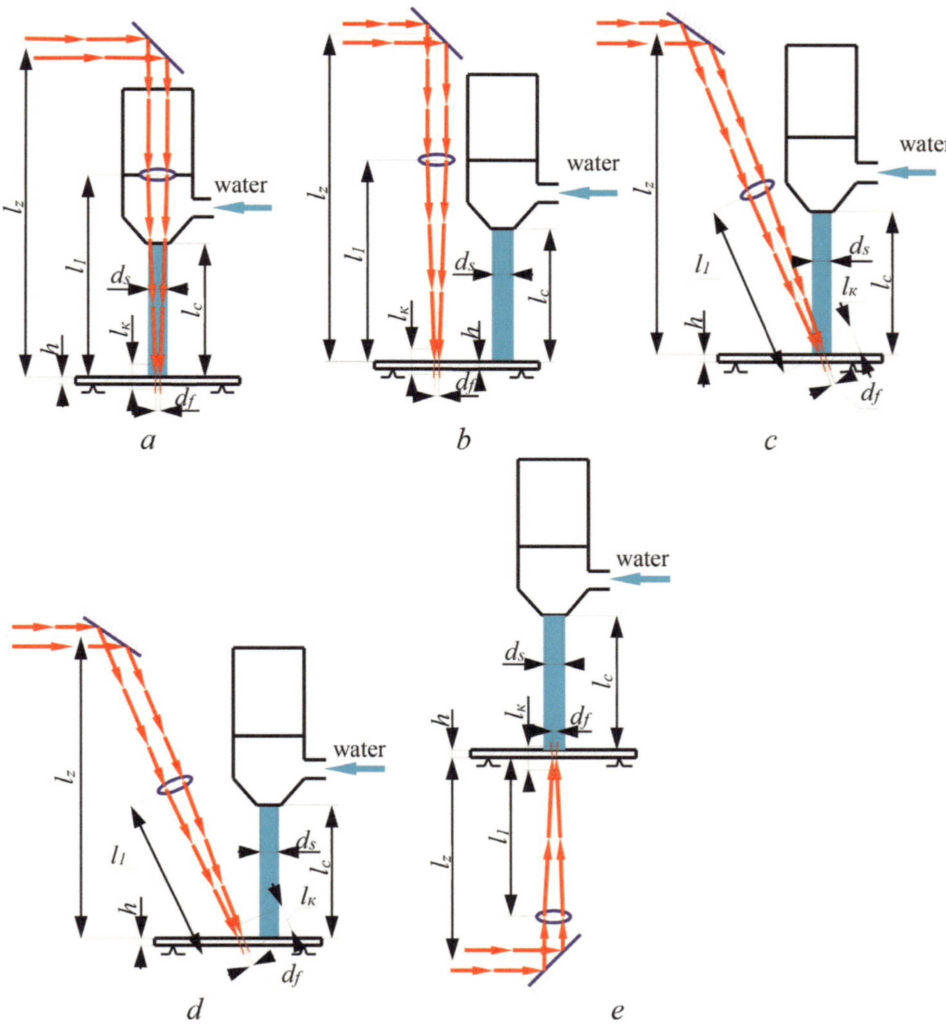

Figure 6.
Variants of combination of a water jet and a laser beam for performance of WJGL (a) or LCC (b–e).

The action of a water jet or a laser beam can be both simultaneous and successive. In this case different variants of laser blasting and possibilities of variation of the form of jet nozzle flow section, as shown in [6], provide good prospects in combination of methods of actions on the worked piece and, consequently, in meeting principle (10).

Application of thermo-hydrodynamic jets typical of WJGL or LCC causes heating of the workpiece whose temperature field can be described by equation [7]

$$T(x,y,z,t) = \frac{P}{\pi\frac{1}{2}\rho c}\int_0^l \frac{e^{\frac{(x-v(t-z))^2}{4\alpha\tau+A^2}\frac{y^2}{4\alpha\tau+B^2}}}{\left[\left(4\alpha\tau+A^2\right)\left(4\alpha\tau+B^2\right)\alpha\tau\right]^{1/2}} \times \times\left[e^{-\frac{z^2}{4\alpha\tau}} - h*(\pi\alpha\tau)^{1/2} erfc\left(\frac{z}{2(\alpha\tau)^{1/2}} + h*(\alpha\tau)^{1/2}\right) \times e^{h*z+h*^2\alpha\tau}\right] dt,$$

where t is the time from the moment of beginning of thermal impulse action, ρ, c, λ, $\alpha = \frac{\lambda}{cp}$, and h^* are density, specific heat capacity, heat conductivity coefficient, temperature conductivity of the workpiece material, and coefficient of heat transfer from the surface of the workpiece, respectively;$\alpha = \frac{\lambda}{cp}$ A and B are the bigger and smaller half axes of beam elliptic section; and $P = \pi qAB$ is the power of the lasing emitter. Integral equation of heat energy balance in an arbitrary area $\omega \subset \Omega$ according to [4] will take the form $\int_w \frac{\partial e}{\partial t} dv = \int_w g dv - \oint_{\partial w}\left(\vec{q_T} + c_g\rho_g T\,\vec{v_f}, \vec{n}\right) ds$.

Taking into account the boundary conditions in the cutting zone

$$c\rho\frac{dT}{dt} - \lambda\Delta T = \frac{(T-R_0)kP}{\pi AB}\exp\times \times\left[-2\left(\frac{(x-vt)^2}{a^2} + \left(\frac{y}{b}\right)^2\right)\right] \times \exp(-kz) \qquad \lambda\frac{dT}{dz}\bigg|_{z=0} = \alpha(T)(T-T_p),$$

temperature distribution across the sample section is obtained; its analysis reveals the following (**Figure** 7): temperature distribution across the surface is determined by conditions of coolant outflow and may considerably vary with the change of the flow shape; when the thickness of the sample increases, considerable reduction of the jet's ability to perform the work of destruction can be observed, as at the same amount of rejected heat, the amount of delivered heat constantly decreases.

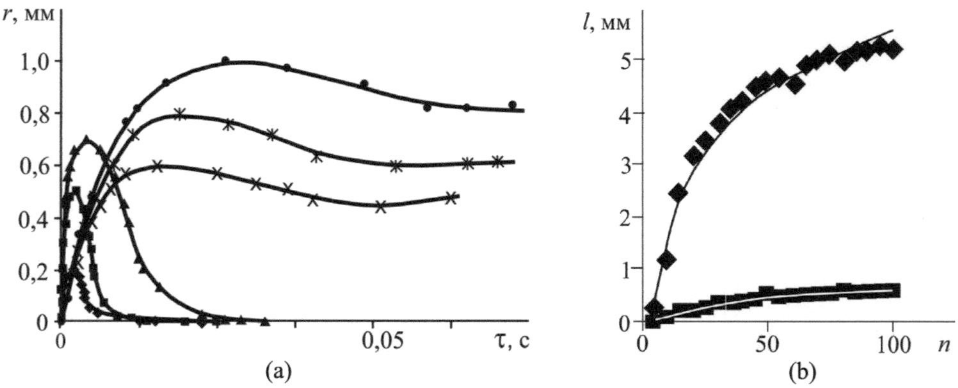

Figure 7.
Changes in border radius zone temperature field (T = 850°C) as a function of time for different processing methods LC (×, Ж, ●), LCC (▲), WJGL (♦, ■) (a) and cutting depth in materials (HA, ♦; PSTM, ■) depending on the number of cuts (b).

Worked material	Functions F'			
	Useful F_{pi}	**Harmful F_{vk}**		**Neutral F_{nj}**
	Obtaining of orthogonal edge of the product (depth h, mm)	**Thermodestruction δ, µm**	**Cracking l, mm**	**Variation of porosity P**
HA	Intensive, linearly depending on t (number of cuts N)	Minimum	Exists	Absent
PSHM	Intensive, decreasing with increase of groove depth h_l	Minimum	Minimum	Is observed in destructed zone
PCD	Nonlinear medium-intensive	Essential	Absent	Is observed in destructed zone

Table 3.
Provision of the product function by creation of element E_i *by a technological action* $W_{ij}(t_k)$*—WJGL.*

Thus, it is possible to create a table of provision of the product with the function via formation of its separate element or a totality of elements. This table is analogous to the one considered above (e.g., for WJGL; **Table 3**), but it reflects the peculiar feature of each of the variants in **Figure 6**. It should be noted that the given tables just illustrate the approach and draw attention to the most important functions, while a totality of such functions may be much bigger.

On the basis of **Table 3**, technological action for obtaining a particular element can be presented as

$$W_{2i}(t_k) = W_1^{Fp1}(t_k) \cap W_1^{Fv1}(t_k) \cap W_1^{Fv2}(t_k) \cap W_1^{n1}(t_k). \quad (14)$$

Corresponding transformations for LB for different materials include four equations now (according to the number of components):

$$\begin{aligned} h|_{W^{Fp}} &= b_0 + b_1 s_k + b_2 T + b_3 Q_v + \ldots; \\ \delta|_{W^{Fv}} &= b_0 + b_1 s_k + b_2 T + b_3 Q_v + b_4 h_l + \ldots, \\ l|_{W^{Fv}} &= b_0 + b_1 s_k + b_2 T + b_3 Q_v + b_4 h_l + \ldots \\ P|_{W^{Fv}} &= b_0 + b_1 s_k + b_2 T + b_3 Q_v + b_4 h_l + \ldots \end{aligned} \quad (15)$$

where T is the impulse energy, Q_v is the liquid discharge out of the nozzle, and h_l is the depth of the groove in the worked piece.

Analogous tables, including the ones with extended features of functions F', are created for all the types of working.

The obtained dependences (12) and (14) make it possible to determine the totality of variants for carrying out the process of working of a particular element of the surface, and the most optimal method can be searched on the basis of morphological analysis.

This approach is applicable not only to the processing of superhard materials. It can also be successfully used for the treatment of special difficult-to-cut materials, such as carbon-carbon composites, multilayer aircraft panels, etc. [9–11].

3. Used equipment, workpieces, and succession of experimental research

Cut workpieces of HA T5K10, PSHM on the basis of CBN "borsinit" represented plates of the size 12.7 × 12.7 mm, of the thickness up to 3.5 mm, DC workpieces were two-layer composites of the size 12.7 × 12.7 mm with an upper

PCD-layer of the thickness up to 1.0 mm and a hard alloy base of the thickness up to 3.0 mm.

The cuts were performed by methods of hydro-abrasive (HAC), laser (LC) cutting, laser cutting with liquid cooling (LCC), and water-jet-guided laser (WJGL).

Experiments were carried out with the use of laser blasting complex LSK-400-5, equipped with a hydro-abrasive head with water nozzle d_c = 0.22 mm and a metering tube D_k = 1.05 mm.

HAC was performed with abrasive consumption up to 0.5 kg/min and under liquid pressure of 250 MPa. Garnet sand of granularity of 300 meshes was used as abrasive during cutting.

During multi-cut laser blasting, the cuts were performed with feed up to 50 mm/min. Impulse Nd:YAG laser with impulse frequency of 75–150 HZ and power of 400 W was used. For realization of the hybrid process, the unit was equipped with a special laser blasting head enabling working with the use of a ring laser nozzle and a mixing chamber to provide hydro-abrasive cutting without disturbance of the position of basic axes of the tools. The beam was focused according to a method providing centering and ingress of all its modes on the focusing lenses of the tube. Optic elements were blown with purified compressed air through a nozzle of ∅ 2.8 mm under excess pressure of 0.05 MPa. During LB the liquid was fed directly into the center of laser focus under the pressure of 25–50 MPa.

A high-temperature carbon-carbon composite weave composite was also used.

4. Discussion of the obtained results

Paper [5] contains comparative results of the research of productivity of different working methods. In the present paper, experiments with workpieces made of HA and PSHM are repeated, but attention is focused on the indices of the quality of the worked surfaces, roughness and accuracy of relative position, and also deviations of their values from theoretically possible ones.

To determine components in **Tables 2** and **3**, the values of separate parameters whose averaged values are given in **Table 4** are determined. Comparison of erosion rates determining the rates of deepening of the cutting dimples is given in **Table 5**. In this case only particular results of the research are shown; they are sufficient for hybridization of the process according to principle (10).

Type of working		**HAC**	**LC**	**LCC**	**WJGL**
Modes of working		100 mm/min, 350 MPa, 0.5 kg/min	30 mm/min, 400 W, 100 Hz	30 mm/min, 400 W, 100 Hz, 0.5 MPa	
		—		2.80 mm	1.05 mm
HA	R_a, μm	12.5	6.3	6.3	3.2
	h, mm/cycle	6.4	5.9	3.5	3.0
	δ, mm/5.0 mm	0.3–0.9	0.02–0.15	0.02–0.09	0.02–0.04
PSHM	R_a, μm	—	6.3–3.2	3.2	2.5–3.2
	h, mm/cycle	$<$0.01	3.0–3.2	2.4	1.5–1.75
	δ, mm/5.0 mm	0.80	0.05–0.10	0.02–0.05	

Table 4.
Accuracy and roughness of the surfaces of workpieces made from HA and PSHM at different jet working methods.

<table>
<tr><th>Type of working</th><th>Modes</th><th>Working by HA (◆)/PSHM (■)</th></tr>
<tr><td>HAC</td><td>p_b = 320 MPa
m_a = 0.6 kg/min.
d_c = 0.21 mm
D_k = 1.05 mm</td><td>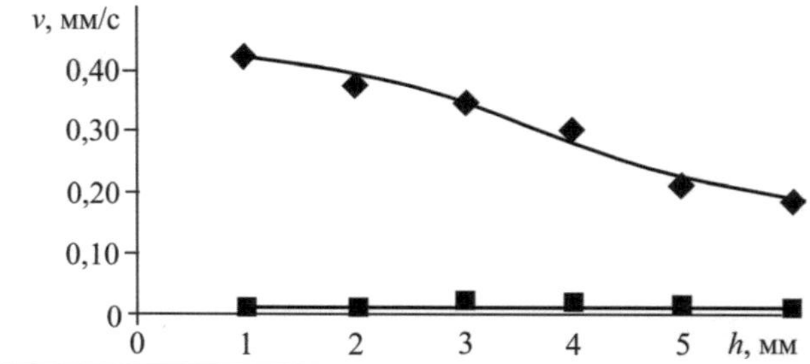
</td></tr>
<tr><td>LC</td><td>P = 400 W
f = 100 Hz</td><td>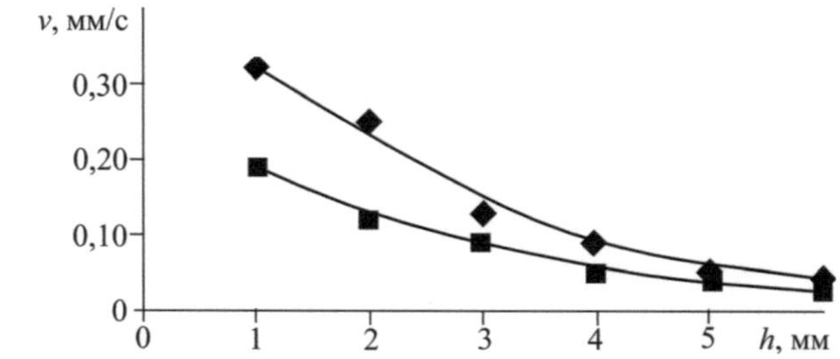
</td></tr>
<tr><td>LCC</td><td>P = 400 W
f = 150 Hz,
p_c = 0.5 MPa
d_c = 2.8 mm</td><td>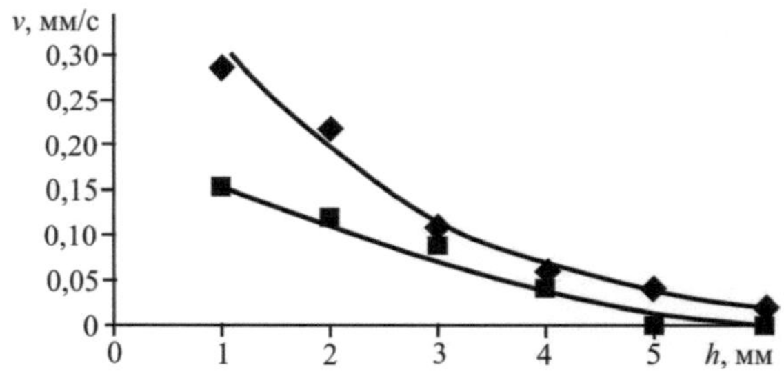
</td></tr>
<tr><td>WJGL</td><td>P = 400 W
f = 150 Hz
p_c = 1.5 MPa
d_c = 1.05 mm</td><td>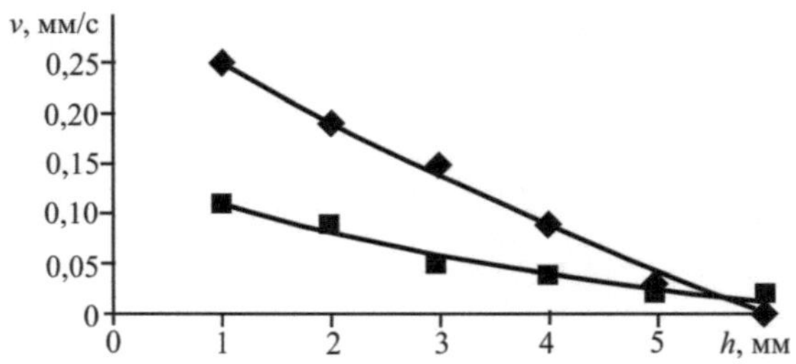
</td></tr>
</table>

Table 5.
Material erosion rate at HAC, LC, LCC, and LB depending on the thickness of the plate.

The performed research revealed the following. Any initial deviation of the cut shape, especially during performance of discontinuous working processes, distorts the jet, and further working results only in deterioration of the quality of the product, first of all, due to progressive deviation of the cut shape. It concerns all the types of working. So, **Figure 8** demonstrates a defect of DC cutting in the form of deviation from orthogonality of the edge at partially incomplete cutting of PCD layer.

A peculiar feature of two-layer DC consists in the fact that workability by a certain method changes on the surface obtained at the section, depending on the structure of the material and physical and mechanical properties. A peculiarity of DC consists in the fact that PCD is not practically worked by HAC, while it can be worked by LCC and WJGL. The carbide base is worked by HAC with high efficiency.

Hence, generation of the surface as an element of the product can be presented in the form of logical multiplication of transformations $W_{1i}(t_k)$ and $W_{2i}(t_k)$ and is described by the expression

$$W_p(t_k) = W_1^{Fp1}(t_k) \cap W_1^{Fv1}(t_k) \cap W_1^{Fv2}(t_k) \cap W_1^{n1}(t_k) \cdot W_1^{Fp1}(t_k) \cap W_1^{Fv1}(t_k)$$

Thus, taking into account the condition of provision of initial high-quality formation of a destruction dimple in PCD layer, a hybrid process is to represent a totality of processes of initial WJGL up to the moment of complete cutting of PCD layer and creation of a dimple in WC with further cutting of the plate by HAC (**Figure 9**).

Further research is to be directed to determine rational combinations of hybridization at multi-cut working, at which the number of cuts directly influences the quality of obtained surfaces, and also to the condition of adhesion surfaces in multilayer products.

Comparison of the diagrams of cutting rates of DC and PSHM demonstrates (**Figure 10**) that productivity of the process is determined by obtaining scribe in the superhard layer. HAC cutting of hard alloy base is practically performed in an identical way.

Let the acquisition of an array of properties of a finished product (a sample of a carbon-carbon composite) in the form of sets Fi (l, b, h, r, δ ...) be possible by the implementation of a certain set of technological influences Mj, inherent in one or another method of processing. Each technological effect can be represented as a

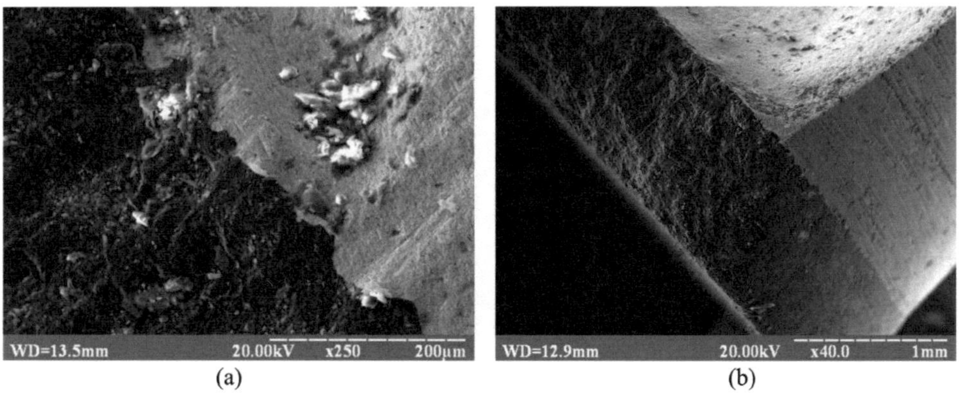

Figure 8.
PCD layer after WJGL (a) and a plane of adhesion of PCD with WC (b).

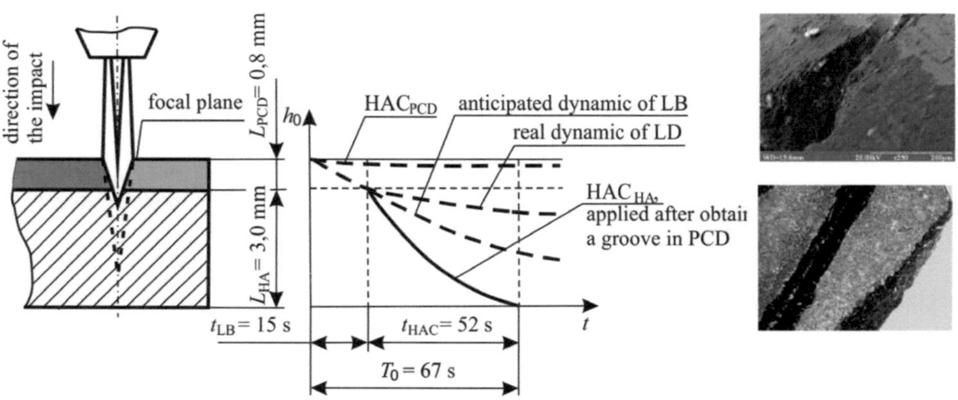

Figure 9.
Making a cut in PCD and PSHM workpieces.

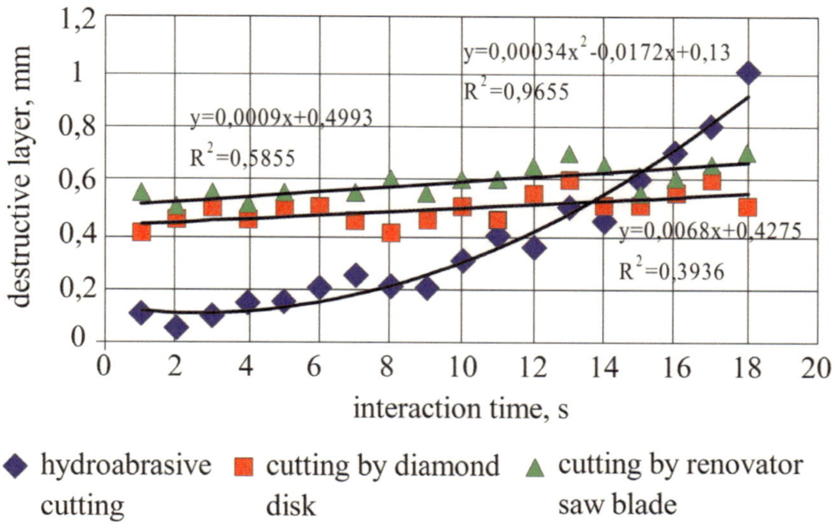

Figure 10.
Change in the width of the destruction zone depending on the time of the interaction of the cutting surface with the tool.

set of sets of properties of the tool Tj, the dynamic properties of the processing system Wj, the method of power influence Pj, for which (Tj, Wj, Pj) ∈ Mj, which gives the opportunity to perform with the properties of Si, output quality parameters to be presented as:

$$Fi\ (l, b, h, r, \delta...) = M_j \cdot S_i.$$

Based on the provisions reflected in the work [8], we can record the condition for providing output processing parameters as

$$Fi\ (l, b, h, r, \delta...) = T_j \cdot S_i \cap W_j \cdot S_i \cap P_j,$$

which gives a set of variants of processing conversions, the best of which can be selected according to a certain criterion.

Since this material is resistant to temperature effects, the only way to treat it is mechanical abrasive cutting.

Take into account the following processing methods: R1, milling; R2, processing with abrasive disc; R3, processing by the renovator (reciprocating motion); R4, drilling; and R5, hydro-abrasive cutting. For these methods, the maximum stresses and diagrams of stress distribution at the intersection of the cut in the workpieces are determined.

The cutting modes, as well as the parameters of the applied tool, directly determine not only the processing performance but also the thickness of the destruction of the surface layer, which must be taken into account when processing samples for mechanical tests.

On the other hand, in the process of processing, there is wear of the tool: the change of angles on the cutting edge and curing abrasive grains. All this leads to the fact that even in stable cutting modes, the parameters of the surface layer are changed.

The change in the width of the destruction zone (**Figure 10**) measured on the natural samples (shown as separate points in the diagram) shows the following. The "hardness" of the hydro-abrasive jet leads to the fact that the power circuit of the interaction is constantly changing and, consequently, the components are increasing, that is, the flow will then simply blur the machined sample, which requires minimization of the time of finding the jet in a stationary state.

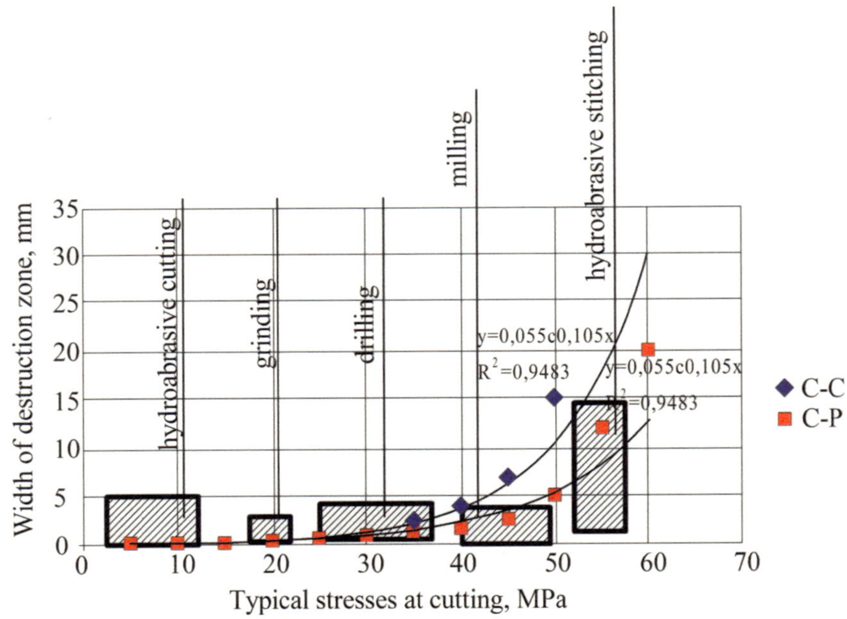

Figure 11.
Emerging stresses under different processing methods and their effect on the width of the destruction zone.

When processing materials with diamond disks or saw blades, there is no significant change in the width of the destructive zone.

We also propose rather simple regressive dependences of the width of the destruction on the function of maximum stresses in the cutting area $\delta = f(\sigma_{max})$:

$$\delta_1 = 0.55\exp^{0.1\sigma_{\max}};\quad \delta_2 = 0.74\exp^{0.08\sigma_{\max}}, \tag{16}$$

where δ_1 is the destruction of carbon-carbon material and δ_1 is the destruction of the carbon polymer, which made it possible to construct and compare diagrams of the development of the width of the destructive zone (**Figure 11**) of the processing of carbon composites, which are formed by various methods.

Thus, we have shown that obtaining qualitative samples of a complex configuration is associated with some complexities due to the heterogeneous structure of

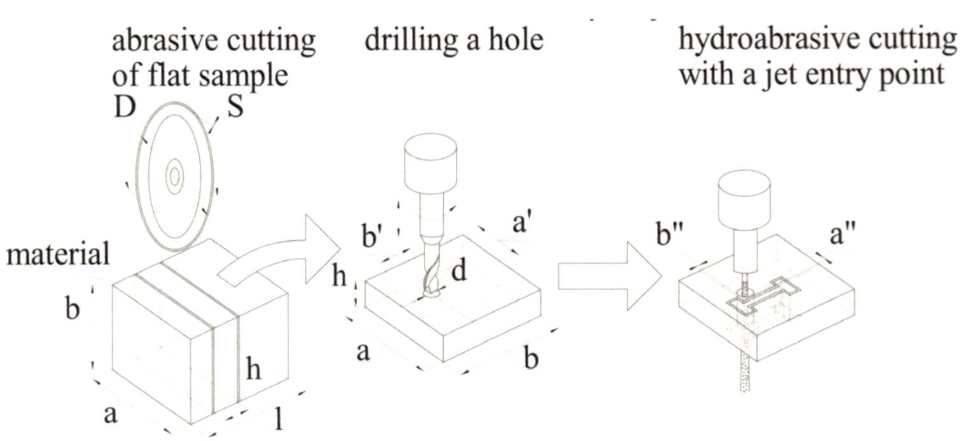

Figure 12.
Sequence of sample preparation from carbon fiber blanket as the result of cutting.

the treated material, as well as the manifestation of quasi-cleaved properties of carbon reinforcing fibers.

In this case, the main condition for obtaining the desired result is the restriction of the thickness of the destructive layer at >2% of the measurement basis.

The application of the proposed approach allows for the task of obtaining a sample $a \times b \times h$ from a cubic blank in A × B × L sizes, to offer the following possible technological operations: R2-R1; R2-R5; and R2-R4-R5. When optimizing the process by the criterion of minimizing the processing time for a given level of quality, a variant of the functionally oriented process (R2)-(R4)-(R5) is obtained, the scheme of which is shown in **Figure 12**.

Such a sequence of operations allows us to make full use of all the benefits of hydro-abrasive cutting (a significant reduction in the processing time) and to avoid material destruction at the time of the breakdown by the jet.

5. Conclusions

Application of the functional approach makes it possible to reveal rational succession of TP steps, find optimal conditions and points of occurrence of harmful functions, eliminate them (or invert), and also determine the possibility of hybridization of the process. Such an approach can be used in working of composite workpieces from laminated superhard materials.

The performed research with the use of HAC, LC, LCC, and WJGL has enabled studying of the intensity of destruction area introduction into the worked piece and finding out the functional features of a particular process that are caused by the working conditions. It has been shown that the hypothesis of quasistationary rate of destruction is unacceptable for such materials, which is caused by the structure and high hardness of the material. It has also been determined that when deepening increases, the rate of jet introduction has a pronounced tendency toward reduction.

A peculiar feature of cutting two-layer composites including a PCD layer based on HA has been analyzed, and it has been shown that the use of functional approach principles makes it possible to propose a hybrid process and corresponding hybrid tool (combining WJGL and HAC), enabling efficient and productive cutting of such plates. Besides, conditions for essential improvement of working process efficiency have been demonstrated (**Figures 13** and **14**).

To improve the considered approach, it is necessary to orientate further research to optimization of working conditions and search for ways of improvement of cut

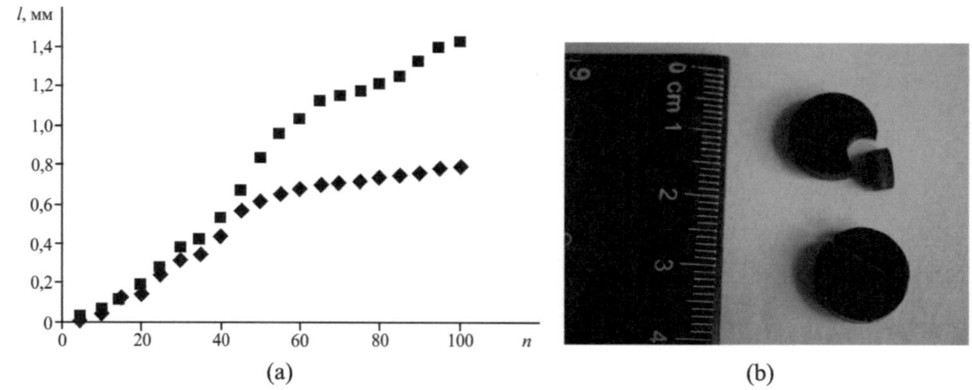

Figure 13.
Intensity of scribe development (a) in workpieces from PCD (♦) and PSHM (■) (linear feed rate, 35 mm/min; impulse passing frequency, 75 Hz) and a cut product (b).

Figure 14.
The result of cutting carbon-carbon composites.

quality by selection of rational parameters of the process and design solutions for the hybrid tool.

Author details

Alexandr Salenko[1]*, Viktor Shchetynin[1], Galina Gabuzian[1], Evgeny Lashko[1], Mohamed R.F. Budar[1], Sergey Klimenko[2] and Alexandr Potapov[3]

1 Kremenchuk Mykhailo Ostrohradskyi National University, Ukraine

2 V. Bakul Institute for Superhard Materials of the National Academy of Sciences of Ukraine, Ukraine

3 Yuzhnoye Design Office, Ukraine

*Address all correspondence to: atmu@meta.ua

References

[1] Novikov NV. Superhard Materials. Obtaining and Application. Vol. 1. Dumka: Nauka; 2003. 320p

[2] Klimenko SA, Mel'niichuk Yu A, Vstovskii GV. Interrelation between the structure parameters, mechanical properties of sprayed materials and the tool life in cutting them. Journal of Superhard Materials. 2008;**30**(2): 115-121

[3] Fedotyev A, Fedotyeva L. The Prospects of Carbolloies Waste Utilization as Wearproof Coverings. Vol. 39. Gabrovo: Izvestiia na Tekhnicheskiia Universitet Gabrovo; 2010. pp. 30-33

[4] Gindin PD. A mathematical model of thermal cleaving of brittle anisotropic materials. Surface. 2010;**1**:14-18

[5] Salenko AF, Shchetinin VT, Fedotyev AN, et al. Methods of cutting for workpieces of hardmetal and cBN-based polycrystalline superhard material. Journal of Superhard Materials. 2015; **37**(4):271-281

[6] Salenko O, Gabuzyan G, Myronov Y, Nikitin V. About some results of processing SiC-microarrays by hydroabrasive precision jet. Journal of Mechanical Engineering NTUU "Kyiv Polytechnic Institute". 2013;**67**:178-184

[7] Schulz W, Niessen M, Eppelt U, Kowalick K. Simulation of laser cutting. In: Dowden J, editor. The Theory of Laser Materials Processing. Vol. 119. Springer; 2009. pp. 21-69

[8] Salenko AF, Shchetinin VT, Fedotyev AN. Improving accuracy of profile hydro-abrasive cutting of plates of hardmetals and superhard materials. Journal of Superhard Materials. 2014; **36**(3):199-207

[9] Kholodnyi V, Salenko O. The water jet guided laser method in punching honeycomb cores for aerospace sandwich panels. Eastern-European Journal of Enterprise Technologies. 2016;**1**(5(79)):19-30. DOI: 10.15587/1729-4061.2016.59870

[10] Ivanova VS, Balankin AS, Bunin IZ, Oksogoiev AA. Synergy and Fractals in Material Science. Moskva: Nauka; 1994. 384p

[11] French MJ. Design principles applied to structural functions of machine components. Journal of Engineering Design. 1992;**3**(3):229-241